COMPLETE GUIDE
TO

MAKING MONEY WITH YOUR IDEAS AND INVENTIONS

the text of this book is printed
on 100% recycled paper

COMPLETE GUIDE TO MAKING MONEY WITH YOUR IDEAS AND INVENTIONS

RICHARD E. PAIGE

BARNES & NOBLE BOOKS
A DIVISION OF HARPER & ROW, PUBLISHERS
New York, Cambridge, Hagerstown,
Philadelphia, San Francisco, London,
Mexico City, São Paulo, Sydney

A hardcover edition of this book is published by Prentice-Hall, Inc. It is here reprinted by arrangement.

First BARNES & NOBLE BOOKS edition published 1976

ISBN: 0-06-463446-9

80 5 4 3

Dedicated to:

THE MAN WITH AN IDEA–

the only creature God endowed with creativity.

WHAT THIS BOOK CAN DO FOR YOU

The *Complete Guide to Making Money with Your Ideas and Inventions* brings you the benefits of basic knowledge which experienced, successful, creative people have accumulated at great cost in time and money, and so, makes it possible for you to build your own success on their solid foundation.

This *Complete Guide* is devoted to making the marketing of ideas and inventions a more scientific, teachable subject, and removing much of the trial-and-error type of thinking from it. By supplying detailed, step-by-step procedures to follow, you can avoid the discouraging results which so many creative people experience. The invaluable concept of actually building success into your ideas and inventions, is one of the most important steps any creative person can learn.

In this *Complete Guide,* you will learn which fields are receptive to outside inventors and which must be for insiders only; you will learn how to judge your own creations impartially, and thus save wasted efforts–and money! Nine major markets for ideas and inventions are described in great detail, and the guidance is such that only a successful and experienced inventor could give you this "behind the scenes" information. Preparing you for your battle is the first essential step toward eventual success.

The *Complete Guide* will bring you factual, selling case histories which show the procedures at work. In Chapter

Seven, for example, you will see how selling your inventive ability instead of a specific invention, can earn for you as much as $18,000 a year and more. In Chapter Eleven, you'll learn the advantages of selling an idea to an established firm to earn royalty checks of thousands of dollars each year, while in Chapter Thirteen, the author tells you how he was able to foresee a need, and with a simple but ingenious locking device, get a United States Patent which he sold for well over $40,000.

Chapter Fourteen guides you to where men of ideas can earn from $15,000 to as much as $50,000 annually.

In this world of ideas and inventions, your success will lie in

SELLING.

The *Complete Guide* will give you a *scientific basis* for planning your success, from choosing the right field for your talents to selling your ideas. Like all other scientific knowledge, the marketing of ideas must be an "open end" subject, changing with the times and conditions of our economy. Only one fact does remain constant: Inventions may differ from can openers to computers, but marketing procedures remain unchanged . . .

YOU MUST SELL TO SUCCEED!

Richard E. Paige

CONTENTS

CHAPTER ONE

HOW TO CHANNEL YOUR CREATIVE THINKING TOWARD IDEAS THAT PAY

Undisciplined creative thinking can be but an enjoyable exercise of the mind, lacking completely in financial rewards. Acceptance and payment for creative thinking can best be achieved by planning the business details first, then the invention; i.e., inventing only what one has the ability to market.

CONTRARY to the many romanticized stories and comforting legends told about successful inventors and other creative people, there is *no* money to be made by creating ideas and inventions; furthermore, there is no money made by being granted a patent or copyright.

The *only* money forthcoming from creative work results from the *selling* of it, not the creating of it, and an inventor's success will be measured by the dollars his sales produce. Success has never been judged by the degree of ingenuity shown—for, until the invention is sold and produced, it is neither seen nor bought by the public.

Obviously, then, to succeed we *must* sell!

Let us go about it as scientifically as we do our creating.

Why You Should Confine Your Ideas to Fields You Understand

The first fundamental step in selling is to ascertain whether the idea or invention under consideration is worthy of the time and expense we shall have to invest in it to consummate a sale or arrange a license. To do this, we must examine not only the idea or invention, but, with equal care, the *source* of the creation.

To try to judge our own ideas and inventions impartially is a most difficult task—to judge ourselves is even more difficult. Still, the success of the whole scientific approach to marketing will depend upon our ability to weigh not only the ideas but the sources of the ideas, and so, it is recommended that we examine our true situations as if we were a third party—fair, probing, analytical and seeking the truth . . . be it pleasant or otherwise.

As a trial project, let us examine a typical situation and observe how the *source* affects the invention. Assume that we, the readers, are this hypothetical "John" and let us examine the *whole* of the problem—*invention and source*—in an effort to judge its worth.

To illustrate: John has a "wonderful idea" for a new brake for an automobile—his sweetheart thought it was "just great"; his mother was most encouraging too. However, John is a stock clerk at a big department store, and though he is most ingenious mechanically, he has not had the background which would have made his judgment valid in such a matter. This does not mean that John's invention is not good, but rather that his lack of a background in that field leads us to doubt his ability to judge the worth of automobile brakes. Therefore, John's idea needs much more work before it is ready to market.

Conversely, had John been an automotive engineer, a gas station mechanic or professional inventor, the idea would be worthy of greater consideration, because we assume he understands his field; hence, he would be less apt to produce an impractical invention through ignorance. Nothing looks quite so brilliant and promising as a new idea seen through the eyes of a person who has only layman knowledge of the matter.

Even with a knowledgeable background there is room for doubt, because we create today in an age of unprecedented industrial activity and our Patent Office bulges with close to 4 million patents.

For the sake of our illustration, let us grant the knowledgeability of our "John." Would this permit us to assume that we are safe and ready to sell? No! We still face such questions as:

Does it really work?
Is it new?
Is it useful?
Is it patentable?
Is it practical to make . . . at a price?
And, most important of all, can we sell it?

If such obstacles face us in a field in which the creator *is* acquainted, what blind groping must we do in a field in which the creator is but an ingenious layman.

To use a hackneyed phrase, first things should come first, and therefore it becomes obvious that an inventor or creator of any salable idea must confine his work to a field in which his knowledge of its needs can, at least, be firsthand. Later, with the detailed guidance found in these pages, we shall learn of all the steps to be taken so that we can follow a proven procedure and avoid finding ourselves promoting some other person's idea, or one inappropriate to our capabilities.

Perhaps the first point to establish, and one that can frequently make the other points needless, is . . .

- Is this merely an idea, not an invention reduced to practice?
- Is this some mental image, sketch, or plan which has not been tried in model form?
- If it is, a model may prove it to be unworkable, and our marketing efforts become inappropriate, or at least premature. No one should confuse an idea, or even a completely drawn plan, with an invention ready to market.

Until you have *an invention,* ready to show to a prospective buyer, ready to demonstrate in convincing fashion, you are not ready to market. Depending upon the field, a "model" could be anything from a workable prototype to a mock-up of the proposed changes. However, *never* is a mere idea to be considered a product ready to be marketed.

This has been and is a very vital point in all marketing plans, for the people who are operating in fields other than their own, too frequently confuse a mere idea or pencil sketch with an invention, and want marketing efforts made for that nebulous, untried, untested and perhaps unworkable "inspiration." Don't do it!

Some creative people are knowledgeable enough to make excellent practical creations for a field in which they are but interested laymen or hobbyists. We must consider their positions: such things can be marketed, but the job is indeed difficult because they have no connections nor prestige in their chosen fields. Unfortunately, too many of our talented creative people operate unscientifically in the "sales areas" of their work, and find themselves with selling jobs to be done which may be beyond their abilities and purses.

No one can deny that a good idea may be sold, but it is unwise to devote effort to a job which is extremely difficult, costly and uncertain. True, we cannot change the past, but we can make the firm decision to *channel creative efforts* into fields we know intimately. Only then will the selling of ideas become a practical matter. If you learn just this one thing from the book, you have received your money's worth.

The younger reader may rightfully ask "What if I have no field?", and that indeed is the plight of many creative people. For an answer, read on.

Why Your Own "Backyard" May Be the Best Place to Begin

While no one can deny that an undertaker invented the first automatic telephone dialing system, and that two concert musicians invented the natural color photographic film, *years* of thought, study, experimentation and work went into the developments before they were ready to launch them commercially. In *basic* new areas of development, there may be no field from which an invention may spring. However, do not be misled. Such inventions were not developed from a background of ignorance. They weren't casual "ideas."

Rules to Follow for Financial Reward

To begin your study of successful, commercialized creation, *avoid the one-in-a-million freak:* instead, choose as your own "backyard" those areas which make up your daily life, be they kitchen activities, minding children, delivering milk, banking, selling insurance or even just going to school. Thomas Edison once said, "There *is* a better way to do it–find it!" And that may well be the cornerstone of your career in creation.

The greatest surprise of your life can be just "learning to see." There are wonders all about you, but they are not to be seen until someone shows you an old thing in a new light–then, never again, will it look old and uninteresting!

There was the story of a toil-worn farmer in the southwest, seeking a bare living from his rocky, barren land.

He gazed about him and thought, "This soil is not fertile. The climate is barely able to sustain crops. What a God-forsaken, worthless piece of land." One day a team of government geologists asked permission to explore his land. He laughed at them and said, "You ought to be able to find a better farm than this to explore, but if you want to waste your time, that's your business–go ahead!"

They did explore the barren land and found that the rocky, barren hills were indeed bad farmland; however, the rocks contained carnotite and other uranium minerals which made the land a fabulous find. Once the farmer learned what he was looking at, it never seemed dull or uninteresting to him again.

The prime requirement for anyone who strives to be an innovator is to see his field with knowledgeable eyes. Far from complaining about his field being dull or having faults, the experienced creative man *seeks* faults, for he has learned that *a disadvantage is an opportunity somebody missed.*

Just examine some of those "dull things," the "nuisances," the "boring jobs"–*those* are opportunities in disguise. What could be duller or more distasteful than changing a baby's diaper? Yet, this inspired the designing of the disposable tissue diaper, which eliminated odor, unsanitary hampers, and laundry bills, while creating a sizable new industry.

Another difficult and boring task was the making of strained foods for infants–a laborious, time-consuming "cross to bear" for all mothers. Yet, Mr. and Mrs. Gerber did not see the unpleasant daily task that other parents saw; instead, they realized in this disadvantage an opportunity to create a huge and prosperous industry–Gerber baby foods. What a difference a viewpoint makes!

No one lacks a field to work in–he lacks only the vision to see it. Whether one works on a delivery truck, in a factory, office or school, there is scope there for an "exercise in ingenuity." Following the old cliché, "Look first in your own backyard," it will be found that while obviously other yards may be "greener," your own still offers one advantage over the others—you understand it better. Build up your mental muscles wherever you are.

Why You Should Invent for Fun Before You Invent for Money

If inventors want to succeed financially in their chosen fields, they must exercise their mentalities as assiduously as a concert pianist would practice his piano. By inventing things for fun, and knowing that he will probably just file the plans or sketches in some drawer or wastebasket, the inventor is not wasting the effort. It is not the plans or sketches which had the value, but *the build-up of inventive experience,* the facile handling of mechanical problems and the familiarity with the subject which has the value. One day this "trial run" inventing will bring forth something of startling value—and you will know it!

Having the patience to follow the step-by-step procedures outlined here will set the stage for success, because only those things will be attempted which have a chance of succeeding.

Seek no guidance here for how to put over an intercontinental missile, a plastic skyscraper or some grandiose harnessing of the sun. Ours shall be a solidly integrated plan for inventing only what we really understand best; attempting only those things which we could finance and, finally, devoting our time to projects which we have the ability to sell or license.

In subsequent chapters, we shall learn why only certain original inventions are worthy of considering as the basis of a new business; we shall discuss why improvements on existing devices are best to license to others in the field; why even though it is called a gimmick or a gadget, an easy-to-make device which really serves a need can not only be patented but often makes more money than a complex machine, and can be the basis for a business. In simple summary, we are aiming to produce inventions with their success "built in" by forethought . . . the kind of forethought which makes an inventor devote his energies to creations which have a very good chance to find acceptance, commercially.

Study the list below carefully, and actually write down on a separate sheet of paper, the questions and your well-considered answers. It will bring out the facts that shall determine the path you take.

Necessary Qualification of Your Idea

Here is a "checklist" that will help you know when you are ready to launch your new project.

1. Is your new idea in a field you really know well?
2. Is there an economic need for your idea?
3. Have you avoided bucking the large, well-established research laboratories and creative staffs?
4. Did *you* search the patent files, the trade papers and the available trade catalogues?
5. Did you actually make a model to be sure your idea was workable?
6. Did you protect yourself with *written* records? Did you have them dated, witnessed and notarized?
7. Did you read the chapter on marketing procedures in this book?

If you can answer "yes" to the above questions, you are ready to take the next steps.

CHAPTER TWO

HOW GOOD IDEAS ARE TURNED INTO DOLLARS

Financial success is not just good luck; it is the product of forethought in the directing of one's efforts. He who would breed whales should not work in a laundry tub. Ingenuity without analytical direction is wasting a blessing.

THE first step in turning your ideas into dollars is to make an accurate appraisal of your situation. Such an appraisal must take into consideration both the invention and yourself, for this information will guide you in your decisions, and having a clear-sighted, unbiased viewpoint will enhance, immeasurably, your chances of success.

Before you try to apply logic and experience to your situation, it is imperative that you have a mental house cleaning to rid your mind of misleading information, romanticized legends and oft-repeated myths, for they could seriously impede your progress. Success requires that you guide your actions with *facts,* not inspirational fiction, and avoid the most dangerous error—the use of outdated, obsolete facts. Yesterday's information may no longer apply; check "facts" periodically. The tales which are supposed to inspire you and offer moral support are not to be taken literally as fact or used as the basis for building a business. Repeating success is indeed impossible—the time is past!

With this in mind, let us examine the inspiring legends and either benefit from their true worth or sweep them out with the other mental refuse as we clear our decks for action. We must indeed know a "fact" when we see one.

Analyzing Success Stories

The keystone of success is recognizing the fact that *each success is a product of its time.*

To illustrate: Charles Goodyear, not a prolific inventor but "a man with an idea," was actually an impoverished tinkerer; he was not a chemist nor a chemical engineer, yet through endless trial and error he discovered how to vulcanize rubber, and so created an industry. He worked at home on his kitchen range and still he made an important, basic contribution to science and manufacturing.

Does this mean that if we had the same mediocre educational background, we could tinker with rubber, or plastics, or chemistry in our kitchens and make a great scientific contribution? *No!* Why? Because Goodyear worked at a time when our whole economy was minuscule by today's standards and technology was indeed primitive.

To make *any* contribution to today's complex, highly developed chemical technology would take a well-trained, knowledgeable chemist.

Ironically, now, even Charles Goodyear would be unable to find employment in a rubber manufacturer's laboratory because of his crude educational background. Chemistry has outgrown tinkerers and amateurs.

The story of Goodyear's eventual success was written to inspire you. However, our own good judgment should recognize the vastly different circumstances which attend chemical technology a century later.

Even knowledgeable chemists have difficulties in opening new areas among the heavily staked out claims in our chemical patents. Layman experimenters should seek new and unbroken ground or they will lose endless days in this maze of knowledge.

Almon B. Strowger, an undertaker with a great mind for mechanics, invented the first automatic telephone switchboard system (1889). Strowger was one of those rare geniuses able to disregard all limitations and bypass all obstacles. As Elbert Hubbard so aptly expressed it, "Fences are made for those who cannot fly," and Strowger's thoughts indeed had wings.

While success in so complex a matter was slow, he did eventually succeed.

Does this prove that undertakers, or salesmen, or mechanics can tackle a telephone invention today? Don't take anyone's advice, just use your own good judgment in weighing the facts. When Strowger invented automatic switching, the telephone art was just beginning to develop and the "state of the art" was primitive.

Contrast that with the almost unbelievable laboratory facilities of A.T. & T., I.T.T., R.C.A. and Western Electric, then look at the vast patent file on the subject—and you will know that tackling such a field as an "outsider" would require either a superman . . . or a fool!

If a man really is a genius (and America is supplied with a surfeit of them), he will prove his mental superiority by knowing to what his thoughts should be devoted. It is such disciplined thinking that forms the philosophy of this book.

Knowing What to Invent

We must never lose sight of the fact that it is more important to know what to invent than to invent it. Choosing the field, the need to be served and the *appropriateness to our capabilities* is 75% of the battle. Stock clerks who invent missiles to Venus are not headed for dollars, they are just enjoying mental gymnastics.

The stories of C.H. Hall and his invention for producing aluminum economically; Edwin H. Land and his Polaroid camera; Clarence Birdseye and his frozen foods; King C. Gillette and his safety razor–all bear out the truth that "big money" from inventions results from *opening new areas of business* and thereby benefiting from the sale of the new *products.* This is a sharp distinction from selling the patents for others to use.

That "big money" came to *those who founded new firms* on their inventions and controlled their own destinies. They were not selling mere improvements in heavily patented fields, nor were they depending upon others to buy their patents and take the financial risks. *Big* risks and *big* money go together.

Does this mean you *must* start a business with your invention? No! It merely stresses the fact that only rarely does any inventor get even moderately "big money" for selling or licensing his patent. He can be well paid, but not in the fabulous figures people tell tales about.

It is here that you must analyze, and be precise in delineating between, inventions worthy of founding a business upon and inventions which may be good and have commercial value, but are not of major importance. These lesser inventions should be sold or licensed, or, if you are well situated, used in your own already-established business. Your ability to see clearly what your invention really is–what it is worth–and how you are situated in being able to market it, will be the keystone to your success. Give it time and thought!

No book can analyze the current situation for you. This book can merely provide you with a viewpoint, a treasury of enlightening experiences and, above all, a fund of information to broaden your outlook and help you arrive at a correct decision on your own. However, it *can* and *does* bring to you the distillation of many successful lifetimes to enrich your (apparently) creative mind, and so will guide you well.

Using Experience as a Tool for Our Task

Wise humans recognize that, looking ahead, the paths open to us are many, varied and lacking in assurance. Looking back, with the perspective given us by the years, we can see why some paths lead to success, others to failure. All humans seem blessed with 20-20 hindsight! However, that same hindsight, while no longer able to influence the decisions for the situations to which they had originally applied, are far from useless.

The right decisions and the wrong ones can now serve as lessons, and they can be tempered with our own judgment, to guide us in our present course of action. We shall drop the sarcastic term "hindsight" and substitute the more useful one "experience." Yes, experience (even someone else's) is indeed a good teacher, but its lessons *cannot* be adopted wholly, for the circumstances which governed the original happening no longer exist.

It is this judicious use of experience which can make it either valuable or downright harmful. Let us extract the useful guidance from historical successes just as we would extract the juice from oranges: we must not think that we must eat this worthy fruit, skin, pulp, seeds and all, to enjoy its benefits.

The Lesson of the Gillette Story

To demonstrate: What can we learn from King C. Gillette and his safety razor? We realize that there existed a need among all men for shaving tools; that all men of Gillette's day had regular straight razors; that most of them found shaving a chore. They had to strop their razor daily, but they did not have to keep buying new blades and then, after a few shaves, throw them away and buy new ones. Because of that, and despite the evident safety and convenience of the safety razor, it was not the overnight success that people of several generations later thought it was.

Had King C. Gillette tried to sell his invention instead of his razors, before a market was established for them, the people who might buy or license his invention would think of it as an *unproven device,* and therefore of limited worth—perhaps a few thousand dollars at most. As a proven device and the basis of a well-established business, the *business* could have commanded a good price.

In analyzing the safety razor's advantages, you find that even though regular straight razors were well known, a *safety* razor was a pioneering invention. Further, it catered to a broad field; it had a big dollar volume potential; it carried repeat business, guaranteed, in the sale of its blades. *If* Gillette could have licensed his patent on a royalty basis, he would have faired better than selling, but still *not* have been a "big money" earner.

Just as royalties usually amount to more than outright cash payments, because the inventor has assumed some of the risk involved, just so, the man who starts a business (whether he is the inventor or buyer) gets the lion's share of earnings because he has taken the greatest financial risk.

Our conclusions from the foregoing story will probably be that Gillette made his money as a razor manufacturer, as a businessman, *not* as an inventor. His invention gave the business a very valuable, exclusive product to sell, and he sold it. He did not merely try to get others to undertake the burden of launching the invention for him.

You will find a very close parallel in the launching of the Polaroid cameras by Edwin Land and countless others: *they* kept the initiative in their own hands; *they* launched the project. Both pioneered new areas of business—they did not merely bring forth an improved version of an old, accepted device. Both benefited by repeat business—film and razors. There are many such successes to point to today, but each one had to be a major development to warrant the investment in time and money. What happens to the inventor with a good idea, yet not one of such enormous dollar potential? Let us go on to his problem.

Getting Your Dollars Through Sale or License

If you have reason not to found a business on your invention, the alternative is to sell your patent or issue licenses under it. If your invention is not patented, a potential buyer or licensee can use your invention without your permission. *Legally, until you have been granted a monopoly by the government, you cannot stop anyone from using your invention.* A wise inventor gets his patent first, *then* tries to sell or license the invention.

If your analysis leads you to the conclusion that sale or license is best for your particular needs, then you will find that you have selling problems much in common with other similarly placed inventors—and by studying their successful solutions to such problems, you may well find the correct methods for handling your invention, despite the fields being completely diverse.

Here is an actual case in which an inventor patented a dressmaker's tool, a pinking shears—today a basic tool in every garment trade, but then a new invention. The problem was that although important, it did have a limited field to cater to, and a business founded upon it would be small, while its selling cost would be great! Also, while simple in appearance, its manufacture was very tricky, and the special cutting tools it required put it in the class of a difficult tool to make and sharpen. While admittedly, a small business could have been established upon this limited-use tool, the costs in manufacturing, selling and advertising may well have made the device unattractively expensive and limited its success. However, the wise inventor realized that as a good "number" in a line of shears, his device could benefit from the sales force calling on the garment shops and cutlery stores, and further, could well use the manufacturing know-how of the shears manufacturer. J. Wiss & Sons, a long-established scissor and shear manufacturer, was chosen for the project, and over the years the Wiss pinking shears became a very worthwhile item.

Frequently, despite a device being a top-notch item to promote, it still may not warrant its own sales force; its own factory; its own company. Yet, as part of some prosperous manufacturer's line, it may be indeed a highly valued money-maker. Judging the correct handling procedures for launching such items can be the difference between success and failure.

It is not just the qualifications of the device which should dictate the procedure to be followed, but also the talents and qualifications of the inventor. He may be a clever mechanic but a poor or completely inexperienced businessman; he may be a good inventor but financially limited to supporting his family, and incapable of setting up a project requiring sizable investment.

There are many licensing arrangements to meet many commercial and industrial needs, and you should acquaint yourself with their well-worked-out provisions, for they may well suggest precisely what you seek, despite their being intended for a different field from your own. As but one of many offbeat arrangements, the author offers the following:

In both the advertising display and packaging fields, the nature of the work requires limited licenses; therefore, the author licensed display and carton manufacturers to submit his patented constructions to specific customers for specific jobs. Most display jobs were one-time uses, and while carton jobs were frequently of longer duration, they too were for limited periods; thus, the trade found it expedient to get permission to submit patented ideas with license extending to that job only.

While so many prospects talk of "exclusive license" arrangements, very few can justify exclusivity financially. The next best thing is "exclusive for their particular customer," regulated by the good judgment of the patentee whose own profit is at stake in issuing conflicting licenses. The licensees were allowed from three to six months to submit their designs and get their hoped-for orders; if this period passed without success, the license expired. A model, submitted for an appropriate fee, was furnished with each license.

Minor Inventions Can Be Major Earners

Many a complex, ingenious engineering feat barely pays for its development, while other simple devices, designed to fill a broad need, pay handsomely. In marketing, we must rate inventions as major or minor by their potential dollar worth; i.e., a vegetable-peeling gadget that could be sold in chain stores and achieve a volume of 5 million peelers annually, warrants greater investment than a scientific device of exceptional ingenuity which caters to a very small, restricted market.

In this book, we are striving to *market* ideas, to turn ideas into dollars–not rate the degree of mechanical ingenuity. We must be sure that we do not let our natural bias for ingenious inventions warp our viewpoint in financial matters.

At this point, let us examine our personal capabilities in order that we may place them upon the scale also, for the weight of position, wealth and prestige must be added to the invention to evaluate its dollar potential properly. If we would rate "ideal" as 10, we should rate each lesser situation proportionately; thus:

- A new device developed by a prosperous man for his own well-established business is the ideal situation; he has the initiative, he has the facilities. Rating: 10.
- A new device developed by a prosperous man capable of organizing a new corporation and knowing how to arrange for financing. Rating: 9.
- A new device developed by a knowledgeable person for his own field, but not his own firm; a most desirable situation, especially if the man has achieved some degree of recognition as an authority in his line. Rating: 8.
- A new device developed by a knowledgeable person who has no prestige to offer, but nevertheless a sound knowledge of the field. Rating: 6.
- Beyond this point the rating is 5 or less, for we are weighting down the invention with many factors, from ignorance of the needs of the field to ignorance of the basics of selling; blindly, then, we give it a 50-50 chance.

As we go beyond this point, the apparent worth of the invention must carry most of the burden, while the inventor's ability to sell adds appreciably to its worth. If you can meet people with confidence, appear knowledgeable and conduct business negotiations as a person of stature, you will find all of these personality factors distinct assets.

Many inventions had to be bought by the self-induced enthusiasm of the prospect, because the inventor was inarticulate or incapable of making a persuasive presentation. Being an ingenious mechanic does not assure a proper reception if that mechanic is unable to speak well. That is why so many mechanically adept, but conversationally inept inventors have chosen one of the following procedures:

1. Working in partnership with a sales-minded man or employing him.
2. Working as an "inside man," employed to design and invent on a salary basis, or possibly for a salary with some royalty or profit bonus.

The vast majority of ingenious men who are unskilled at sales and negotiation, are best situated as valued employees. They thereby bypass the perpetual marketing problems. You may find that you *can* sell yet prefer being a company employee, for you have greater scope for your work and greater financial backing. Many a good inventor has become a highly paid corporate officer.

Analyzing the Two Creative Types

While creative minds actually vary too widely both in quality and volume of their creations to be classified accurately,for the purposes of this book, we shall divide them into broad classifications:

1. The prolific, naturally creative person, usually happiest in a creative job.
2. The man with only occasional ideas, or perhaps even but one. Usually, he is a businessman who can devote himself to promotion of the idea rather than continual creation.

All too often, vanity will cause a man with a good idea to pose as a highly creative type, when actually some of the most successful inventors were "one-idea men"; i.e., Maytag with his one idea . . . the Maytag washing machine. Building his *business* around the washer was his form of creativity. The professional inventors, the prolific idea men, being interested in creating rather than selling, "lay their eggs and walk away from them," because it was the joy of creation which they enjoyed—*not* turning the ideas into dollars through business.

A classic example of the prolific inventor was Thomas A. Edison, with his never ending productivity. Still, he realized relatively little from his enormous output of truly basic inventions. Several businessmen made tens and even hundreds of millions on his electric lights alone. Edison, after a lifetime of producing basic things, such as the phonograph, electric light, motion pictures, poured-concrete houses, etc., died leaving only $5 million. His mind was mechanically creative; he had little interest in business, and suffered much discomfort because of it.

If we seek financial remuneration, we must rate ideas by their dollar potential, not their ingenuity or complexity. Much has changed since Edison's day, and with a world grown heavy with complexity, patents, financial structures, and fast-disappearing frontiers, today's success—you, I hope—must select *possible* things to do, things which you believe you can *market* . . . not just invent! Many a well-marketed, simple, inexpensive device has made someone rich, and many a truly fantastic computer, with fantastic engineering problems as well, has been too unwieldy to market.

The subsequent chapters of this book will be devoted to the actual step-by-step details of the art of marketing inventions. Broad enough categories of ideas will be chosen to show you that while fields do vary widely, selling methods have certain basic patterns to follow. With ideas—big or little, important or trivial—*selling them* will be the key to success!

Classifying Your Creative Type

For your own guidance, write down on a separate sheet of paper the questions found here and your frank answers, so that you may properly evaluate your position.

1. Am I a professional, prolific idea man?

2. Am I an occasionally creative person?
3. Are my ideas important financially?
4. Are my ideas basic and original enough to become the foundation for a new business?
5. Am I the inventor of an improvement in an already-established art or field?
6. Would my invention, if manufactured, justify a sales force of its own, or must it join a product line of related articles?
7. Can I handle, adequately, business correspondence, meetings, negotiations, financing, directing others?
8. Would I do better financially as an employee, selling both my inventions and my services?
9. Should I seek a business partner or employ one?
10. Should I risk my own capital to start a business, or should I seek outside investment capital?

When you have given yourself straightforward, frank answers to these questions, you will see more clearly where you now stand and what steps must be taken to reach your goal.

CHAPTER THREE

HOW TO FIND A COMPANY TO MARKET YOUR IDEA OR INVENTION

Experience can teach us the scope of the authority of those with whom we must deal; adjust to their limitations if you would have your selling efforts succeed.

THIS chapter sets forth the detailed procedures to be followed by those inventors who desire to *sell or license* their patents to others. Chapter Six will cover founding a business.

The great preponderance of inventions are either improvements upon or extensions of existing devices. Such devices are usually best suited to sale or license (except when capable of exceptional marketing performance). If your invention falls within this category, the first step should be to consult with a reputable patent attorney; have a search made; then evaluate the worth of what you have, both as an invention and as an item to market.

While, obviously, a patent attorney is not an expert in each field his clients see fit to engage in, after a search is made, his opinion of the invention's market value and potential sale may well be worth seeking.

If you decide to try to market your patent, either by

sale or license, then you must learn the procedures which will make possible the acceptance of your creation, on your terms. Here, it is of prime importance to bring out one fact: while *inventions differ widely, selling methods are similar.* To demonstrate the truth of this, the author has chosen three examples in completely unrelated fields, and goes into the details of trying to sell or license the inventions in the way you may well have to attempt yourself. You can learn much by studying these examples, from first selection of prospects to final closing.

In this chapter, you will learn:

1. What kind of company to select and why.
2. Which member of the company to choose for your proposed presentation.
3. How to set up a meeting with that executive.

In Chapter Four, you will learn:

4. What to discuss, what to show, what to reveal, at your presentation meeting.
5. How to compose appropriate financial arrangements and write contracts.
6. How to perfect your sales techniques.

These six procedures will furnish you with sufficient information to allow you to handle your arrangements efficiently.

Selling vs. Producing

It is *not* an oversight that the subject of production methods, manufacturing facilities and factory problems have been entirely omitted. This was done deliberately because, only in the *production of merchandise* does production precede selling. In the *selling or licensing of inventions, SELLING PRECEDES.*

As an inventor, you would be wiser to devote your

energies to making business contacts and selling your creation, than planning how to manufacture your eventual product in some as yet unknown factory. At such time as production planning is in order, some specific production facilities, with men to run them, will be brought into the picture, and *they* will have ideas to place at your disposal. *Selling* is the problem which needs your time, ingenuity and planning *now!*

Most creative people find production planning more enjoyable than "knocking on doors," selling and making presentations; however, those who do make the effort find it most rewarding. It is almost a form of "industrial acting" before a select audience, and if you discipline yourself to do it, you soon perfect a rare and valuable technique—you become an "entrepreneur" who can turn an invention into dollars . . . an art that will stand you in good stead for the rest of your life.

There are times when failure teaches us more than success in the developing of dependable selling techniques; moves we make and things we say which produce undesirable results can be avoided in subsequent presentations. This allows us to work up a very effective sales presentation devoid of such errors. As long as you have to pay for your failures, benefit by them.

Now, let us get down to cases! We shall take the three subjects which comprise Chapter Three and apply them to selling three completely unrelated inventions for the purpose of demonstrating that *all* selling of *all* inventions must follow certain procedures, despite the difference in field, physical form and type of prospective customer.

For our examples, let us examine:

1. A new electrical household appliance.
2. A new automotive accessory.
3. A new way to make brassieres.

As we take the necessary steps to market these three items,

we shall try to find a *common procedure* which all must follow to open the doors of a prospective corporate buyer or licensee, contact one of its chief executives as our proper audience and set up a presentation meeting with the proper authorities. When we do find the procedures which work well for us, they will become part of a treasury of scientific techniques we shall collect and use to take the guesswork out of selling inventions. In every scientific research, *bits* of knowledge must be found, tested and finally adopted or discarded. When you find something worthy of being added to your collection, you will have another valuable tool to work with—another asset to make *you* a more valuable salesman.

What Kind of Company to Choose for Your Prospect

Selecting a company to whom you will make a presentation may seem a simple decision; however, it most definitely is not! Frequently, the fault lies in the inventor misinterpreting his needs. The most common fallacy which guides the action of inventors is they seek manufacturers who can *make* their devices. What they should be seeking are manufacturers who could *sell* their devices.

Applying the Technique to the Marketing of a "Juicer"

In the first example we chose, this point becomes apparent. Let us start with our hypothetical kitchen appliance; to make it typical of its type, we shall give it these features: we have invented a new, ultra-simple, electric-powered orange juice extractor . . . a "juicer" to the trade. Let us say it has a novel mechanism for extracting the juice and is very simple to clean after use. We have applied for a patent, had claims granted, but the patent did not issue yet . . . all typical circumstances.

With this typical situation, we shall try to select a corporation which would seem well-suited to our needs. Most inventors would be quick to point out, with pride in their thinking, that they must not approach a company already having a good juicer, for that firm's juicer would be in direct competition to theirs. They would fear that the company with the product already on the market would either buy their idea cheap for the express purpose of shelving it and killing competition, or just turn it down cold.

In contrast to this, the inventor would point to a fine machine shop with all the necessary manufacturing facilities, but no competitive juicer. What an opportunity this would seem to be—to start him in the household appliance business . . . with the new juicer as his first item. That shop does look like a "natural," doesn't it? . . . or does it? Sorry! No, it is not! Why? Because:

1. Only a manufacturer with long experience in the housewares field could launch such an item.
2. Only an experienced man (and company) could judge its worth and have the volume of sales necessary to produce it at the right price . . . in competition with other juicers. The inexperienced man would have a very high price because he could not have the most essential volume of sales (having no established sales force).
3. The surest road to failure is to start a new, untried product with a new, inexperienced firm. It is not easy to launch a new product *with* an experienced housewares firm; if the seller and the product are both untried and unproven, that is a bad team indeed.
4. A company with *a line of products,* all selling to a housewares buyer, even though the products are not necessarily related, can spread the cost of its sales force over that *line,* not burden *one* product with the entire expense of selling. *Only* a major item with a well-established following can do that and still stay an attractive commercial project for the salesmen.

If the chosen manufacturer with the good factory facilities but no established sales force in housewares wanted to "take on" the new juicer, then he could possibly turn to certain independent sales agents who carry many items, and have sales representation that way. It is *not* desirable for a company to devote a lot of sales time to your invention if it might turn out to be limited and undependable. A manufacturer would sooner handle a line of products than one lone product, even if it seems an excellent item.

It would appear, after examining the facts, that a housewares manufacturer who has already demonstrated an ability to make, sell and promote household appliances would be our best choice.

Knowing that such a firm already has adequate facilities for design and development of new items, we come to realize that only if our item is exceptional in design or price, can we hope to sell the firm on the idea of taking on our invention. However, if it *is* just a moderately workmanlike juicer and has no exceptional features, then we should be working in the firm's design department or in the corresponding department of some industrial design group who cater to such manufacturers. There, some top designer thinks up the basic new ideas, and the lesser-lights work it out and make it practical to manufacture.

You may persist, as does each new inventor, by saying, "This company is our greatest competitor; how can we risk going to it?" The answer is this:

> No matter how good or how successful a juicer they may have, *the TRADE will demand new ideas* and new models each year, or, at the most, each two years. If your idea is outstanding, it could well be just what they are looking for.

If your idea has good mechanical features, but your model is crude and not smartly styled, the company may give it to their industrial designers to "smarten up" and beautify. If the mechanical ingenuity is worth adopting, the styling is added to it by firms or individuals who specialize in that phase of the work.

In choosing a firm for your presentation, you can get an idea of the market by visiting any large department store and noting the various manufacturers whose labels appear on the type of merchandise in which you are interested; however, before you do choose one, you should get one or more copies of the housewares trade publication, *Merchandising Week*, especially at the time preceding the housewares show. Several others are good too, but for the purposes of choosing a prospect, *Merchandising Week* is more than adequate.

This magazine is published by Billboard Publications, Inc. at 165 W. 46th St., New York City 10036. Single copies are 75¢; subscriptions are $15 to an "outsider," $10 to people in the trade.

When you have gone through the issue you bought carefully and selected *several* likely prospects, go to the main public library. There you should consult the *Standard and Poor's Directory* or the *Dun and Bradstreet Directory* for such data as the approximate sales volume of the company, the number of their employees, their main office address, the names of the various officers (which will be discussed in the next chapter), and with that to guide you, you can select a prospect of the kind and size you desire. Remember that the smaller (but still large) companies *are more dependent on outside designers.* Also, you may find you can contact the decision-makers, not the administrators. Now, let us move on to the next phase of our fact-finding.

Which Member of the Firm You Should Select for Your Approach

You will soon learn that each company has its own

internal set-up for handling new-product submissions. Some, with giant facilities of their own, hesitate to even examine any outside submission because they themselves have so much in the course of development that they fear being sued for using something which had been shown to them, when they had been working on it before the submission. Such companies insist on considering a patented item only (or will compromise on a patent application), but they do not guarantee any special consideration for your having gone to the expense of a patent or application. It is safety for them, only.

The various levels of size and worth of the companies dictates the formality of the submission procedure. Even a smaller company may ask you to sign a release (which you should do) to promise not to sue them for having examined your submission. Your patent protects *you* despite the release; if they had your idea before, they would have to prove it before the Patent Office or a court of law. Don't walk in fear or you will never sell your invention. Now, in choosing the man to see, a General Electric or Westinghouse would *tell you* whom to see, and even if your juicer was run by perpetual motion, you could *not* ask to see, nor be allowed to see, the president.

The industrial giants give you a regular printed form or booklet laying down the rules for your submission. If you want to deal with them, you must follow it. The next level down, the large but not gigantic firms, usually have New Products Departments, which find, develop and evaluate new products . . . like yours or *their own.* Parallel to this New Products Department is a Research and Development Department (called R & D) which *does* develop new items and also improves and works on the adopted products of outsiders like yourself. A great portion of the *basically new* ideas are produced on the outside–the run-of-the-mill good ideas may be born in R & D Labs.

In the New Products Department, you can go no

higher up than the head of such a group; no company officer would "go over their heads" to see you. Don't waste your time trying. That in-company facility costs the company a great deal, and therefore they *want* to use it. The company officials expect that good ideas will come from the New Products and the R & D Groups, and unless your idea was something they were unlikely to get from their own men, they would not buy it. Just being good is not enough—*they* are supposed to be good. If you are really outstanding, you may make it.

Now we come to the companies whose sales volume ranges from $5 million to $25 million dollars annually. These are my favorites. They are not so small that they cannot properly finance the new products they see fit to launch, nor are they lacking in facilities for clever design. However, like the difference between a man who owns a car and a man who hires a taxi, the smaller firm prefers to retain top-notch industrial designers *when* they are needed, as opposed to maintaining giant, permanent design staffs year 'round.

Here, in a medium-size, but prosperous firm, the owner or partners enter the development of new products, sometimes working with some man or men who (for prestige sake) call themselves a New Design Department, New Products Department or even an R & D Lab. However, the owner or partners keep a tight hand on the sort of work their new products staff is allowed to do, and no new submission of merit gets by without being seen by the president or vice-president. Remember, however, that not all vice-presidents are owners; many have this prestige title with no ownership nor financial authority.

The benefit of dealing with the owners lies in their ability to make decisions, write checks, enter contracts and become friendly with a promising inventor. In many large companies today, the "turnover" in help is distressing. In smaller houses, the owner is constant. Even top executives of the giant companies are only employees and lack decision-making powers on their own. Had you been favored with a good reception by the head of either the New Products or the R & D of the larger company, you would find that upon their decision to go further with your idea, they would bring in a corporate officer. He would then discuss finances, and, if you seemed reasonable, would negotiate for a price or royalty percentage. You could *not* have contacted him to begin with; he appears for the financial arrangements of a deal whose sales aspects have already passed the New Products or R & D Group's particular authority.

While the *Standard and Poor's Directory* provides the names of top personnel, you will find the heads of departments left to direct inquiry–either a letter or phone call is best. Always get the person's full name, official title, department title and any other details which insure your approach as not appearing careless and casual. If the books you need are not in your library, ask your banker where they may be consulted. Large companies have them, mail-order companies use them for their mailings and *main* branches of libraries get copies–buy them for reference.

Applying the Technique to the Marketing of an Automobile Accessory

Let us now set up another hypothetical item to market to see how our methods apply. This will be an automotive accessory which we will endow with certain useful characteristics and which will also be in the patent-applied-for stage. Let us say that the accessory is a device which attaches under the dashboard; it can be set by a simple stem and knob so that when the motorist exceeds the speed limit, the device emits a warning buzz which continues until the speed selected for that area is reduced . . . thus avoiding getting a ticket.

Past experience has proven that inventors head for

the biggest firms like homing pigeons . . . first! True, a General Motors or a Ford could incorporate such a device in their regular cars; however, what is the situation as it relates to unrequested submissions? Well, first we find that such companies do not even encourage independent inventors because they have such enormous facilities of their own; independent inventors would necessitate their revealing what they themselves are developing, which they will not do. Automotive development is so complex today that an outsider could not begin to appreciate the size and extent of it without having made it his life's work. However, we have our device, and we now shall attempt to sell or license it.

It should be mentioned, as an "aside" that certain fields have become a "patent jungle"–a hopeless maze of interlocking patents known in the trade as "patent structure," which no layman-inventor should tackle. Thomas A. Edison was fond of saying that it took more intelligence to know *what* to invent than how to invent it. The automotive field is indeed such a jungle. However, we do have our hypothetical automotive invention to market, so let us rush in "where angels fear to tread" . . . just to see if there *is* a way open to us.

There is no doubt that General Motors, Ford and Chrysler, upon your written request, will send you a pamphlet of "rules" you must observe if you would have them agree to a presentation. Those rules have to do with having a patented item, so that on no pretext will you be "showing them something in confidence." They do allow those inventors, conforming to those particular rules, to have a polite, receptive audience–your chance to sell the Big Three. If you have a truly basic, world-shaking invention, you might do it, too. Short of that, however, how can you sell to the largest, best-staffed research laboratory in the country? Your efforts could be compared to peddling homemade chocolate bars in Hershey, Pennsylvania.

It should be stated that certain *recognized authorities*

in various specialized fields *are* used for special purposes; this does *not* mean that a layman-inventor can be equally welcomed into their midst, for his contribution would have only the slightest chance of not being a duplicate of some work their own vast force is undertaking.

Once again we recall the quotation that there are more roads than one into Rome . . . or General Motors. How? *Many* devices, large and small, find their way into General Motors' products–devices initiated and introduced elsewhere by the myriad suppliers who serve the giant motor manufacturers. There are an amazing number of such suppliers for the simple reason that several suppliers serve them on each need, and there are needs from floor mats to engine blocks, spark plugs to tail lights. This does *not* mean that the suppliers are an easy path to riches. Quite the contrary–they need *exceptional* things to get business from the Big Three.

While for an unrecognized inventor the Big Three seem a waste of time, the suppliers are at least *possible,* and therefore, to go on doggedly, trying to sell coals to New Castle or chocolate bars in Hershey is not showing "character" but rather stupidity–don't do it! Do not try to be a superman; police your creative thoughts; devote your time to "doable" things–and make money. Study *why* the suppliers succeed while you fail.

They are not selling inventions . . . or patents!

They are selling "*merchandise*" . . . parts . . . devices . . . things! Maybe you can make them *your* things.

Each of the suppliers has a reputation for his own particular specialty, and therefore his word carries weight with the motor car manufacturers. *They* still test *all* devices, but with the layman's suggestions, they never come to a test. Suppliers are easier to approach as a general rule because, unlike G.M., Ford or Chrysler, they do not have a large research and engineering staff–they are more makers than innovators. Like all generalities, this too must be modified according to the suppliers you choose; that is why it is necessary to do a careful job in locating, classifying and understanding the function of each firm you select.

While many pieces of "optional equipment" are available to the motorist from his car's manufacturer, a tremendous variety of items are marketed through automobile stores, Sears stores and even gas station display rooms. Many manufacturers bypass having to sell to the Big Three in favor of going direct. This gives a greater "base" to their sales, and they are less in danger of losing an account with a huge volume, all in one place. Many auto gadgets are marketed in these auto stores, being sold from colorful displays and display packages, and catering to the car-bugs who are always looking for something to add to their most proud possessions . . . their cars.

While many of the dozens of suppliers to the Big Three are indeed minuscule beside G.M.'s $50 billion annual sales volume, do not rate them as small or even modest in size, for a great number are publicly owned, stock-on-the-market corporations who cannot be approached casually under the mistaken notion that being "only suppliers" they are "small and personal." That is where you must once more go back to your *Standard and Poor's Directory* or the *Dun and Bradstreet Directory* and study it for the firm's size, number of employees and corporate officers. If there is a single owner, or just a few partners, it will be more to your advantage than if you see that there is a formidable line-up of officers, especially in Research and Development and Engineering.

Just as many modest-size firms are supplying certain of the needs of the Big Three, in much the same way, firms like Du Pont, Libby-Owens-Ford and Inland Steel are also "suppliers." Study your market with great care; go to the large-city automobile shows where you see all the devices, gadgets, improvements and cars—and you will soon know whom you should approach with your idea. Also, the manufacturers of gas-saving devices and other specialties, such as our speed control indicator, use the magazines like *Popular Science, Popular Mechanics* and the *many* motor car publications. Between these sources of names you will locate your most suitable prospects, and, with the *Directories,* you will get the pertinent details.

So, to summarize what is being recommended in this overly large field, seek out manufacturers who show the kind of sales enterprise you need for your device; select one in a field akin to yours so that you may benefit from his sales facilities (do *not* try to teach him a new business) and finally, take the trouble to learn as much about him as possible, from his sales volume to his personnel, through poring over the *Directories.* Learn to give prospects opportunities they are already set up to take advantage of, with the least amount of changeover.

Marketing a Brassiere

Now, we go on to our third project—the brassieres—and see what we have learned and how it may possibly apply to this entirely foreign merchandise. To go to an extreme to illustrate, the writer has chosen softgoods instead of hardware, ladies fashion apparel instead of mechanical devices and a field where the usual "industrial approach" appropriate to a machine project might be out of place. The *one* point we must never lose sight of is: we are not selling brassieres—we are selling *inventions,* and with that point of similarity, our procedures *will be* similar.

To add credence to our hypothetical brassiere manufacturing project, let us set up our "invention" thus: Because we observed that brassieres are made with breast-fitting cups of graduated sizes, and those cups are made of several sewn-together sectors or sections, they cannot be made to collapse flat for packaging. In the case of the higher-priced brassieres, the garments are frequently wrapped and stuffed with tissue to hold the cup shape on display. Being higher priced, the salespeople can devote the time to them. In the medium-priced, largest-volume line of brassieres, the garments are folded and inserted into a relatively flat carton. When the brassieres fold down, they wrinkle and muss, and do not come out of the cartons looking as crisp and freshly pressed as manufacturers would like.

Our "invention" is a way of cutting and piecing the form-fitting cups, whereby they are as hemispherical as the regular cups when being worn, but when flattened for ironing (laundering) or for packing in the carton, the cups fold on such lines that they lie flat . . . without wrinkles.

The brassiere field is really large—in fact, impressively so. Many women wear brassieres, and some prosperous, well-dressed women have many, ranging from everyday utility models to fancy models to wear with low-cut evening gowns. The higher-priced models are for department store trade, and in contrast, the medium-priced and very low-priced models are, today, packaged in colorful cartons and displayed on a rack, ready to pick up and pay for—no trying on or guessing about the fit, style or quality. This is where the volume lies, and this is where a flat-folding brassiere which will look well after being folded would be most desirable to sales.

While we reserve the right to backtrack and change our minds *if* our experience shows us that starting with the better-grade garments is wise, we can but assume that those who pack the big volume of medium-priced brassieres and suffer the ill-fitting, wrinkled results of flat-packing, would welcome our invention and be prepared to pay well for it.

Here, to locate the firms doing the best job in packaged goods, it may be best to make a tour of the big stores—the medium-priced, large-volume outlets—for such goods. Don't bother with the very exclusive shops, nor the custom-made departments—keep to the departments featuring the packaged brassieres in greatest profusion. In this project, it may be assumed that the inventor is already conversant with all the brassieres so packaged. However, for the sake of clarity, let us go over the ground again.

It would be best to buy samples of the various brands to take home and study, and also collect the names of appropriate manufacturers who sell the grade and kind of brassieres we seek to change. The magazine section of the *New York Times,* Sunday edition, often shows undergarment and brassiere ads: *Vogue, Harper's Bazaar, Seventeen* and other magazines of this type are better than trade papers for our purpose. Once again we must consult our two directories to ascertain the size, sales volume and personnel of the companies.

It may be true that the ladies' undergarment industry has no giants comparable to General Motors or General Electric, yet, for the size and nature of the trade, you will be pleasantly surprised at the size and apparent prosperity of the firms involved. In that medium-priced field for which we are headed, as an instance, we find that one company, Maidenform, Inc., has a large factory right in lower Manhattan, two others in Puerto Rico and a huge office force that occupied an entire large office building on lower Madison Avenue. This size suggests that we examine carefully whom we want to approach.

Such companies have designers, cutters and manufacturing personnel, but do not require the Research and Development type of set-up we encounter in mechanical merchandise fields. Such companies are not beset with the type of patent structure building that G.M. or Ford would. They have problems, too, but their products have more to do with design patents than mechanical patents.

We shall find that there is *not* a prescribed routine of men to see, as we found in the automotive and electrical appliance fields. Our best direction seems to be to write directly to the president, saying, as before, that we have a new development in the manufacture of brassieres which could be of great value to him, and we wish to make an appointment to present it, at his convenience. If he wishes to have certain of his associates present also, you will be pleased to have them. Keep the initiative, as before, by saying you will telephone for an appointment on Tuesday (or whenever you desire) at 10:00 a.m. and hope that the meeting can be arranged. Allow him a day or more before calling.

If *he* wishes to suggest other more appropriate officers of the company to make your presentation to, let him do it, but do *not* work through a purchasing agent, for you have no *goods* to sell. Be gentle, firm and, of course, polite, but get your way on this last point.

As one of the things to learn, be sure you acquire the correct nomenclature, so no misunderstanding takes place. As an example, "lingerie" is not brassieres, "ready-to-wear" refers to dresses and coats, and even though, literally, brassieres *are* ready to wear, the trade does not refer to them in that manner. Consult the trade papers for a knowledge of your prospect's trade, so that you will know enough about it not to appear ignorant before the selected executives. Follow the same routines we found best in the first two projects—write . . . don't drop in; *you* telephone for the appointment; gently hold to the top-level men for the showing; confirm by letter; get there *early*.

One last word before moving on to the second part of this chapter: There is one ever-present pitfall that accompanies the low-echelon presentations; it is the situation in which the inventor assumes that if the presentation is well received at the lower level, he then will be invited to present it to top management. Companies do insist upon "screening" of ideas because so many impractical things are offered that they would waste the precious time of the top executives. However, when the subject matter is found to be of sufficient interest to the company to warrant consideration, then the *INVENTOR should be allowed to show it to management.*

Many a low-echelon man will tell the inventor that his

idea is good, and he–the low-management man–will present it "upstairs." Under *no* conditions must this be allowed. *You can* get a group presentation for an important development. He will not. He would casually walk up to the factory manager, sales manager or other less than top man and nonchalantly lay your idea on his desk saying, "Some man walked in with this. What do you think of it?" No proper staging, no prestige, no impressive presentation with reasons and examples–nothing, just . . . "A man walked in with this . . . " Many an inventor has thought his invention failed to get their interest, when in reality the *man who presented it failed.* If anyone should fail . . . *you* fail, after you do your best and see that everything has been done right. Do not *assume* that your contact man presented it right and the idea was not good enough–*never assume.*

Reaching the Decision-Maker

The very keystone of success in selling or licensing your patent lies in presenting it effectively to a person capable of buying or licensing it–in simplest terms, selecting wisely the potential prospects to whom you intend to sell.

Far too often, creative people would rather "relieve themselves" of their pent-up enthusiasms by pouring forth their stories to people available to them, rather than policing their emotions, suffering delays and doing the work involved in . . .

1. Determining the *right* company to go to.
2. Determining the *right* man to see at that company.
3. Arranging a meeting with *him,* and no other.
4. Refusing to be shunted off to a person without authority.
5. Going to another prospective company rather than wasting time with those who could not possibly deal with you.

Inexperienced inventors think that meeting with purchasing agents, factory foremen or other financially impotent individuals is preferable to having no meetings at all. Granted, inventors are mere humans and may need the therapeutic value of "getting things off their chests," but, if you do this, *know* that you are merely relieving yourself–do not confuse it with marketing your invention. Success can come only if you *hold out for the right man.*

What to Do When Your Letters Do Not Produce Results

In Chapter Two, you read how to arrange an appointment and were cautioned never to "just drop in." There was mention, also, of the inadvisability of disclosing the invention in a letter intended only for making an appointment. Should this letter fail to bring the desired response within ten working days, then further steps must be taken.

In the writer's experience, such letters, if carefully written, do produce a reply; however, if the two weeks pass without such a reply, telephone the company; ask for the man's secretary and, in a nice, friendly way, explain that you wrote to Mr. ______________, and received no reply. Then ask, "Was that letter received?" "Was that letter read?" If you find that he did glance over it, but buried it under "more urgent matters," ask the secretary to please find your letter and place it before her employer, with a note saying that you will telephone two days later for an appointment.

Company heads *do* travel; they *do* have other pressing matters; they *do* tend to sidetrack anything that may relieve them of unnecessary work. Do not let your feelings be hurt, but remember: *The squeaking wheel gets the grease* . . . so bring *gentle* pressure to bear. Remember too . . . keep the initiative!

CHAPTER FOUR

HOW TO PRESENT YOUR IDEA OR INVENTION TO PROSPECTIVE BACKERS

A single lifetime cannot encompass all the firsthand experience necessary to high achievement; we must discerningly collect the accumulated knowledge of our predecessors, and, using it as a foundation, build to new heights.

CONTINUING with the inventions of the electrical appliance, the automobile accessory and foldable brassiere, let us now take the steps which lead us into actual selling efforts. The previous steps were for the purpose of giving our hypothetical inventors their "big chance" to present their creations to people carefully selected for their ability to buy or take a license to manufacture and sell the inventions we have to market.

We have finally arranged our appointment for a presentation meeting before top management. Having taken so much trouble to contact the right parties, we cannot have less than the best thinking of which we are capable, to make the meeting produce the results we set as our goal—acceptance and financial arrangements with our prospect. Would it not be wise, therefore, to learn: 1. what to show; 2. what to discuss; 3. what to reveal—so that we learn *in advance* what would be best; what would be necessary; what would be harmful—and, above all, that we have come prepared, in a knowledgeable manner, to give the basic information a prospect is entitled to receive. Conversely, we must recognize, *in our preparations,* what information may be harmful to our own best interests . . . and withhold it.

What to Show at a Presentation Meeting

Let us begin with one of the more common forms of presentation, which so many inventors and designers love, and which can lead to failure . . . fast!

The "Pretty Picture" Presentation

When an inventor arrives for his presentation meeting, armed with a portfolio of beautiful sketches, smart layouts and handsome color schemes, he may get admiration for his artwork, but his invention is stillborn—it's dead!

Just showing a "wonderful *idea*" for squeezing oranges, with a new and simpler method, and showing a pretty housewife posed squeezing the oranges with a smile on her face, will not stop top management from asking, "Haven't you built a working model?" If your answer is "No," you have said, in effect: this is just an idea—I haven't even tried it out to know if it really works—yet I have the nerve to come here and take your time to consider buying it. Some amateurs even go further; they ask the customer to build the model for them. Moral: go properly prepared, or don't go.

The very least an inventor owes the prospective buyer, is to offer him an invention which has been built, in model form, thoroughly tried out and brought to the point where it is worthy of being considered "ready for market" as a new item. Past this point there will be days of work necessary, but if our host at the management table doesn't think the idea was proven practical by demonstrable use, he will never sustain the necessary interest to take an undeveloped invention and have his own people work on it to make up for the inventor's indifference.

Some inventors think they have overcome the mistrust of the "pretty picture" presentation by, instead, showing very carefully drawn blueprints or plans. However, they soon find out that the pictures, at least, could be understood at a glance, even though they were not believed. The plans would take study on the part of a prospect, and *until he is interested in the invention,* he would not devote the time to study the blueprints. The *working model* is an absolute must for a demonstration of how well a device works, and a well-drawn blueprint could contain innumerable unworkable parts which no one could spot, without hours of patient study. Don't try to make your prospects do your work for you.

One further caution: never say that you could not afford to build a model because it cost too much. That is your own private concern–if you cannot afford a model, don't go into the business.

Another presentation method, used by industrial design students, is a miscarriage of a good idea. I refer to the mock-up in clay, balsa wood or paper board. To show *styling* in three dimensional form, with color and form that can be studied well, the mock-up is excellent–but *not* for demonstrating a *working* machine. The beautiful 3-D dummy will prove a complete dud, as will any other excuse for not having a working model to prove what it will do.

Presentation of a homemade-looking *working* model is indeed the standard practice. Firms are used to these early versions with no dress-up, and find that they actually have the advantage of allowing the works to be studied; whereas, the finished-looking models with their coverall fronts, may look smart, but do not serve the main purpose . . . to evaluate the working parts.

If our new device for extracting juice is indeed novel and worthy of being promoted, then the styling can be added. Not so, conversely; no businessman buys a smart "dress" and then tries to get a machine to place under the decorative cover. There is one further advantage to a working model: if anything is wrong or even unworkable, you can find it out yourself and correct it.

To demonstrate our juicer adequately, a dependable working model must be built. This may be a simple brass or steel framework, onto which has been mounted, in a workmanlike manner, the motor, switch, juice-extracting parts and, if possible, the "easy to clean" parts we claim as our feature–perhaps, made in ceramic or in some easy-to-carve plastic to simulate it.

The actual demonstration of how it works should thus be *shown,* with an explanation, in as many words, pointing out the features. It should *not* be an explanation in words spoken over an inanimate dummy machine or colored sketch. We must give our host credit for recognizing ingenuity and quality in a machine, despite its lack of smart styling. However, if we do have a dummy mock-up for styling, *plus* the unadorned working model, we indeed have a first-class presentation.

We must always have the patience to practice doing all the things we intend to demonstrate, until we can perform them with ease. There is no excuse for practicing before our audience, nor finding ourselves with something we forgot or which doesn't function, when we should have learned of that and corrected it in private. Also, prior practice often turns up many a worthwhile improvement because we learn what the new design needs to function better. We must "live with our ideas" every day, until we really get to be authorities on the subjects and are worth listening to.

Frequently, a great help in making a working model is to purchase one or more of the competing devices on the market and "cannibalize" their parts, using their bases, motors, switches, main shafts, framework, bowls or other less characteristic parts, so that the resultant model has adopted the more common mechanical elements but does not get to look like the competitive machine. Construct only those parts which make the new machine distinctive. Those who have attended industrial design classes have learned many tricks of the trade in achieving smart-looking results with a minimum of dollars and a maximum of ingenuity in buying available, ready-made parts and adding desired features as if they were indeed part of them.

If we are to demonstrate how oranges are squeezed, we would best make a neat, *small* kit, which might contain: four *juice* oranges, a knife, two small glasses, several paper towels, a plastic leakproof bag and a simple plastic cloth to lay upon the conference table. Above all . . . take a 10-foot extension cord, lest we find ourselves having to give a demonstration in some inconvenient corner, just to be near an electric outlet. In essence, we must run through our demonstration in practice, to help foresee what we shall need, and hence, arrive for the demonstration, which really counts . . . prepared!

Thus, in our most careful manner, we try to give our idea every chance to "sell itself" under the best conditions and to the people capable of buying it. Now, before we leave this phase of our activity, let us apply what we have learned to the other two items.

In a manner similar to the way we prepared to *demonstrate* the juicer, we should prepare to demonstrate the speed control device. In this case, it becomes necessary to show the device in working model form *on a car* . . . probably installed in our own car, but, lacking that, installed on a borrowed car. The persons to whom such devices are shown are used to road demonstrations and expect them. However, it is necessary that, besides the working model which will show the idea in action, a second model (also a working model) be neatly mounted on a board, to allow the prospective buyer to study the actions without disassembling the car model. The dummy mock-up for styling is necessary for appearance, and doubly so as a substitute for a working model.

Going on now to our fold-flat brassiere, we see that some half-dozen models should be made available for a meeting. About two or three should be folded perfectly and installed in paper boxes to demonstrate how neatly they fold for packaging (one may be in a clear plastic box so that the buyer can *see* the brassiere without opening the paper boxes to peek in). Also, an inexpensive store fixture of a plastic dummy-form of a woman should be used (bust only, of course), such as specialty shops display on their counters, on which we shall show the brassiere fitted to the dummy, proving that our new patterns for cutting not only lie flat in packing but fit perfectly for wearing. Finally, it would be wise to have two models to handle while talking, without either undressing our dummy or emptying our boxes.

Throughout these procedures, there emerges a theme for our sales efforts–a basic, simple concept, which is . . .

Don't just "talk a good invention,"

rather

"Let a good invention talk!"

Now, let us go further, to check on what we may say at a meeting to help or harm our project.

What to Discuss at a Meeting

When we make our appointment for a meeting, we should ask if we may come 20 minutes earlier to unpack and set up our materials in the conference room. While most fair-to large-size companies do have conference rooms, if more modest facilities are provided, we must do our best to make a formal presentation under whatever circumstances are made available to us.

We must try to set up the meeting wherein, subtly, *we* can conduct the proceedings, once we are past the first formalities. After all . . . it *is* our "show." Setting up where we can face our audience in *one* direction is best; having to turn first left then right as each subject is handled is annoying to *all* concerned. Professional speakers watch that carefully.

Here is a most necessary bit of sales psychology to be learned and used immediately:

A. *Before* showing our models we have a *brief* story to tell, and we must present it in its logical, natural sequence—no plunging into last things first.

1. Why we think there is a need for a better orange juicer.
2. Why we think that the market for a simpler, cheaper, better type of device, like ours, is still wide open.
3. Why we invented a juicer to meet those needs—to break away from all juicers using the same mechanics.
4. How we simplified mechanics, reduced manufacturing costs, eliminated hand assemblies, etc., to get price advantage.
5. How we featured the one thing housewives want—easier cleaning and reassembly after washing.

During this important time in our presentation, we do *not* want our invention seen. Why? Because if it lay exposed before them, our audience would not be listening to us talk . . . they would be busy looking over the model. Our points would be lost. *Keep the model covered!*

When we are ready to remove the covering (anything from a simple piece of cloth to a neat corrugated box), we place the working model before them . . . simply . . . *no excuses,* such as "This is just a rough model" . . . *no excuses at all*! Excuses are the sign of an amateur. Give them credit for knowing the true state of affairs. You may say, "This model shows the mechanics out in the open so you can study our new method of extracting juice." A coverall to dress up the rough model on display merely becomes an obstacle in the way of their inspection. There will be plenty of time for giving it more eye appeal later.

We shall move *slowly*, but confidently, and run the device "dry" as we point out the features. We deliberately take our time, knowing that they never saw it before, and we want them to have plenty of chance to look . . . to let the machine sell them . . . to let *them* sell themselves. So, quietly and unobtrusively we talk, providing "background music" as they soak up the details.

We must avoid the nervous, too quick, "slam, bang, one, two, three!" kind of delivery which satisfies no one but the speaker. Equally bad, and to be avoided, is the attempt to estimate prices for those far more conversant than we are with such details. We may wish to show price advantage, but can do so, authoritatively, only by contrasting labor and materials–not by dollars. Avoid misstatement; be conservative; inspire confidence! That should be our motto. Flamboyance is fine . . . for sideshows.

Now seems the time to say, "I will show you how this new device actually extracts the juice of the orange," and we proceed to take one of our precut oranges and, with no fuss, squeeze it into one of our small glasses. Then, just to add a little fillip of appropriate showmanship, place a second glass to catch the drippings while we hand the glass to our host to drink.

We ask the others present if they too would like some juice, and this helps the demonstration along. Considerately, we keep the waste in our plastic bag, and with our paper towels, show how easily we can wipe clean our touted "easy to clean" parts and reconnect them to the working model again. All the time that the discussion and demonstration goes on, the juicer is allowed to stand before them–clean–for their perusal.

At *this point*, if we have a smartly styled dummy mock-up to show, it can be an asset. We showed the "works" working, and now we show that our working model can be prettied up. Of course we may leave our work open to criticism from those who think us poor stylists, but at least we are not selling the styling. Yes, *with* the working model a dress-up is an asset–alone, it is worthless!

Comments should flow at this point, (Comments from the host or his associates should flow *without* our asking for them.) However, if the people have reason to withhold their comments, and we wish to know where we stand, we *must use a positive approach.* As an example, *never* say, "Didn't you like it Mr. Jones?" but instead . . .

> "I believe this easy-to-clean juicer will have great appeal for the housewife. Don't you, Mr. Jones?"
>
> or
>
> "An advertising copywriter could have a field day with these features Mr. Jones."

Salesmen are notoriously afraid of silences. They ascribe to them many, *untrue,* assumed meanings. Mr. Jones may be cagey, and although impressed by our show, hesitates to show his enthusiasm for fear the price will go up. We may squirm under his silence, but, nevertheless, let us show our mettle. Time enough to think it a rejection when Mr. Jones comes out flatfooted and says, "Sorry! I must turn you down. It is good, but we are developing an idea I think will be even better." If that is the verdict, on to a second showing . . . with us even better prepared.

However, we may just hit him right. Why not? We *do* have a good device–we did pick the right company, the right man–and we said the right things. That *is* what success is made of. He may ask for time to think it over. If he does, thank him sincerely for his patience and go home to write him a note, saying that you appreciate his giving your device his careful attention and you hope he may see fit to adopt (or license) your device soon. *No* mention of royalty, purchase price or *any* financial details until we are specifically *asked for them.* Such details are out of place at a showing. No one is interested in what *we* want, until they decide that *they want* our device. We must not "jump the gun."

To quote a sensible, if not grammatical philosopher, "Patience is what you should have when it don't do no good not to."

What to Reveal and What to Avoid

By some compelling perversity, many inventors feel it incumbent upon them to point out the faults of their own devices. It is better to stress the good points and defend the faults, *if they are brought out by others.* Every person and every device has some degree of imperfection, but at a meeting whose sole purpose is to sell or license our device, such tactics are, to say the least, unwise.

Our broad policy in conducting sales meetings should be to confine our talk to those things which are necessary, appropriate, beneficial and, above all, constructive. If some member of the group sees fit to ask personal questions or request information which should be kept confidential, that person must be handled in a sophisticated manner–refused, but nicely.

If a man from the company's R & D group says so ill-advised a thing as, "We have been working along similar lines to yours. When did you first start to work on yours?", our reply should be: "We never need worry about a point like that, Mr. ______________ ; we both should leave matters like that to the impartial examiners in the United States Patent Office." While such situations are rare, they do occur, and we must keep alert to the fact that dates of conception, reduction to practice and filing dates are strictly our private affair; and knowledgeable people in the field do not feel free to casually ask about them any more than they would about how much rent we pay or what we earn. Those dates could affect an interference proceedings or a court case some day and must be saved till then. You will note that your own patent attorney does not ask you for such information except in wanting it for some actual case.

Another circumstance to review, merely to avoid being caught unprepared, is the too-aggressive company head who demands, "Did you peddle this to all my competitors first and get turned down?" (Beware of the misplaced sense of humor which strangers endow with too much seriousness.) For such a person, the answer may be, "I am here to sell it to you, Mr. Jones. I cannot discuss any other prospect with you. I agree to withhold showing it further until you give me your decision."

If after a meeting our prospect asks that the model be left with him for further study, leave it. We have already taken all the steps necessary for our protection, so, regardless of the possible dangers of copying, stealing or other skullduggery, we need have no fears. Our only fear should be, to have the device remain *un*sold. Businessmen of intelligence do not steal; they can well afford to pay for whatever they want. We are protected by *knowing* what we had to do *before* we came here.

Next to be considered is our problem of protection for our prospect's firm. Here, we come to realize that we should offer our device for sale or license *only* when some generic claim *has been allowed.* Why? Because we are offering our prospect a monopoly; however, until we have been granted one, we do *not* have it.

This most important fact should be emphasized because, *we* do not need a patent to give us the right to make, use and sell our invention *if* it does not infringe any other patent. Obtaining a patent is only for the purpose of *excluding others* from doing so. Had we offered our novel juicer with no patent applied for nor granted, our prospect would be entitled *under the law* to use it, *without your permission.* It is in the public domain–our "gift" to the public. Protection is up to us.

Even operating under a pending patent is not full protection, although it is done often to get a "head start" on the market. Delay can cause some items to die, unborn, for *no* market waits for any inventor–if a need exists, and you can't fill it . . . someone else will! The deciding factor usually is, the *cost of setting up to manufacture.* A short-lived novelty with inexpensive steel rule dies says "Go!". A complex injection-molded item with a $25,000 to $50,000 mold cost says "Wait!". It is understandable.

To illustrate the advisability of taking a chance in certain fields but not in others, let us consider some novel notion display which could have a short but profitable life in supermarkets and retail stores. While it is new it can sweep the field, and in two years be a dead issue (though, if good, may be revived years later). If the manufacturer of the display has a thorough search made, he is relatively safe in not waiting out the patent grant time.

In contrast to this, a manufacturer of hotel and restaurant equipment, who may have developed a new device for automatically broiling steaks to specified degrees of rareness, so that large parties and banquets can be handled by but a few chefs, would not be wise to sell such costly equipment without having claims granted, and preferably the patent actually granted, before selling his devices to the caterers. The display motors would be used and gone in a few months; the cooking equipment would be there . . . vulnerable to suit or seizure in case of an infringement. Further, it would cost more to sue the display manufacturer than the case could bring. The automatic steak broiler entails an elaborate manufacturing set-up, and taking chances with that is foolhardy.

When a legitimate prospect wants to study your application to appraise the allowed claims and judge their worth before he makes an offer for the patent, he should be allowed to do so. He cannot upset the allowed claims nor can he change the status of the application, but if he is wise, he may have his patent attorney work with your patent attorney toward broadening your claims and even finding important things you may have overlooked. In one patent the writer had sold, despite his great care, he had left himself most vulnerable and did not know it. It was one of the early six-can carriers. Previous carriers required elaborate turning in of tabs and locks to hold in the end cans. The writer invented a carrier where the cans actually pushed open the locks, which then swung back and locked the cans in. The need for greatly increased production speeds at the brewer's production lines had been left unsatisfied by the existing carriers of that day.

The writer had a "pair of swinging-door tabs" which swung back under pressure of the can insertion; it worked very well. However the lawyers saw a *single* swinging-door model I had thought not as good and hence did not cover in the claims. The attorneys insisted on covering the single tab, like it or not. In the huge success of the device, *only* the single tab was used; my two-tab models would have left me unprotected, for the claims *all* would have recited the "*two* swinging doors."

A knowledgeable firm can finance a reissue, if needed; add to the claims; send the attorney to see the examiner–all the things "too costly" for you to do. Knowledge *and* money really help!

How to Shape Financial Arrangements

When our demonstration has created sufficient interest in those who attended our meeting, the proper member of the firm will broach the subject of financial arrangements. *Only* the proper man would bring it up. The usual wording for the question we shall hear would run approximately thus "What have you considered asking for your idea?". . . informal, low key, casual.

Don't turn the question around and ask him to make an offer; it is not advisable for two reasons. First, it is our project, and he knows we must have given it much thought. He did not. Second, we don't want the embarrassment of starting with a low or ill-considered figure and fighting our way back up. No! We do *not* want his offer. *We* will give him a well-considered figure, with a little added for bargaining.

While it is highly unlikely that one set figure would be appropriate to prospects of all sizes, a leisurely (luncheon) talk with our patent attorney will evolve in our minds, a *generally* acceptable *level* of pricing, if not a set price. There is far more guidance available to you than you may suspect, and the writer, with many negotiations behind him, now knows that other people's problems were sufficiently close to his to have helped set him right. Thus, we had our luncheon meeting *before* our demonstration meeting so that we did not allow the enthusiasm of the prospects to cool while we ran over to our attorneys' offices and conferred. We did it first.

What did our attorneys tell us? Well, their background showed them that for every outright sale of an invention, there were 100 or more royalty agreements; for every sale of an invention which never before was on the market, there were dozens of inventions which had proven themselves–made good on the market–hence, they established a known worth. (Many such inventions were sold as a valuable asset of a going business, thus commanding an even greater price.) Naturally, the still-to-be-proven invention, *no matter how promising,* cannot command the level of price which either of the two categories can. Thus, discussion clarifies our thought and modifies our position.

When our host does ask us for a figure (*if* we wish to *sell*), then we should be prepared to suggest a price. There is a rule in business—and it goes from cars to diamonds, company acquisitions to complex machinery—*price is open to negotiation.* Maybe a suit or tie, a box of candy or can of paint is price-fixed, but everything else, from an IBM typewriter to a house in the country, is open to negotiation. In Europe, it is a "sport"—a game of skill. Here, we seem to regard it as an unpleasant task. The smoothness and casual handling of the matter are indeed assets to the bargainer. *Watch carefully for a cue from your host* as to when, where and how he wants to negotiate . . . most likely *without* his lesser-lights around.

CHAPTER FIVE

HOW TO BENEFIT FROM A REJECTION OF YOUR IDEA OR INVENTION

There is more to be learned from failure than from success; yet, no one welcomes failure. Good judgment results from bad judgment, benefiting a person who is observant and intelligent.

No one is surprised to learn that to be an engineer, doctor, lawyer or pharmacist, one must go to college and study hard. Yet, most inventors are of the opinion that, uneducated in the complexities of mechanics, patent law, contracts, selling to top management and learning the methods of financing, nevertheless they can launch an invention on nothing more than a mixture of hope, enthusiasm and energetic groping. They cannot! Even *with* adequate preparation, it is difficult—without it, they are wasting time!

Perhaps the most important step in learning how to market inventions is to keep the project moving forward—despite delays . . . despite stupid treatment . . . despite "doing the right thing" but producing no results. The one lesson which this author seeks to teach is: *learn* from each disappointment and use that hard-earned knowledge on another, more skillful try. Look upon your tries as "schoolwork" which eventually will turn out a "graduate" invention marketer, who *will* succeed.

How to Evaluate a "Logical" Turndown

When a supposedly good device is shown to a most knowledgeable prospect and that prospect explains to you why the invention is no good–he *may* be right! Perhaps you had best chuck the whole project and discount your losses. However, the author has learned certain amazing truths: *all* men–even the most intelligent–are *limited in vision*, are actually applying *their* needs and *their* viewpoint to something and finding it lacking . . . *for them!* They are right . . . *for them!*

You could take a nice woman who you think is most desirable and find 100 men who would *not* marry her, who would think her too fat . . . too thin . . . too short . . . too tall; yet, she could find *some man* who would give her a lifetime of happiness. She doesn't need that 100, provided she has the strength and vision to find Mr. 101. A "turndown" may mean that you are selling to the wrong type of prospect; hence, you had better re-examine your sales plan.

Never abandon a project on one man's opinion. Try other men, other companies, other fields. You must look upon this painful sort of "education" as eventually valuable. So long as you are learning, you cannot lose. Remember that.

How to Proceed After an Apparent "Preconception"

Perhaps nothing is as disconcerting and distressing as making a presentation to a prospect, only to be told, "We had that very idea and found it no good." or "We did that very thing years ago. You can't get a patent on that!"–and then, try to go on. *Yet, it can be done.* How? By mentally going through that very scene, and variations of it, and, at your leisure, thinking out the best answers . . . something you would not be able to do "under shock" if you did not learn this procedure.

The author *has* gone through this several times and found that, while it *is* possible that someone could actually be correct and *maybe* stop you cold, it is 95% possible that you still can go on and succeed in spite of the so-called "preconception." *Learn these answers . . .*

1. Many people had "an idea" and abandoned it. In the eyes of the Patent Office, *you* gave the benefit of it to the American Public; you deserve the patent. Of course, someone else's public sale and use *could* stop you. Get details *if* possible.
2. Many an idea which was expensive to produce, failed during the depressed 1930's. In a newer form, launched during the later, more prosperous years, it was again "new" and succeeded. Do *not* give any snap judgments on the spot. Study the evidence; then come back to meet again.
3. In a lifetime of experience, the author has found that most people's idea of "identical" is so inaccurate that it can be discounted, and it *should be* until *proven–meanwhile, suggest that the meeting proceed.* Word of mouth is *not* proof: even a picture in a catalogue may be unacceptable to the patent examiners, for pictures do *not* show mechanics–a dead man and a sleeping man would look the same in a photograph. An actual model *may* be proof.

Always remember that *time* is your ally–do not make hasty moves nor commit yourself on the spot. Get details if you can: *have good arguments prepared* and go on calmly. He who goes on with courage usually readjusts and succeeds.

When some person does have an identical invention to show you, even then do not give in. He may have had it for years hoping to "someday" do something with it. In the eyes of the Patent Office, *he abandoned it*–you did not. He may have conceived it before you but did not reduce it to practice till much later than you, in which case you would still win. Study the situation, acquire patent ingenuity–it pays.

How to Prevent Being Sent to a Person with No Authority

If the man to whom you have written tries to send you to his purchasing agent, his factory foreman or sales manager, you must handle him artfully. *If* he wrote suggesting the switch, you answer in letter form, too. If he telephones or speaks to you when you call, you must tell him *the matter you wish to see him about requires a man with authority to negotiate financial matters*–and that cannot be anyone but a member of the firm. You will be pleased to have the man he suggested present at the meeting, but you cannot show an important invention to a person without authority.

Unless you are a person of most unimpressive appearance and manner, he, or some vice-president or partner, will see you. Sometimes a second letter seems in order; this is especially true when the man you wrote to was away on an extended tour and you cannot depend upon the secretary to dig out your letter and show it after all that time. If the president handles his affairs in so shoddy a manner that he does not even give you the courtesy of a reply, then he would indeed be an undesirable man to deal with–look elsewhere. There are plenty of better, more considerate people to deal with. Find them! Remember, you can't sell everybody. You need only one good prospect–keep trying.

Develop the attitude that you are showing a president of a company something to increase his business; that no one is too important nor too busy to see an important new development and, finally, that the only elements in your talks which are to be forgotten are those concerning what *you* want. What your own selfish interests are *cannot be pertinent* to the situation until the man or company shows interest in the project.

How to Handle a Mild, Polite Reception

It has truly been said that if you do not know whether you want to marry a woman–don't! Why? Because if you are not in love with her to the point where you really *want her,* skip it. Similarly, if your letter-phone-call technique gains for you a meeting with the right people, but the result is a mild reception–a polite interest but no real enthusiasm–then just thank them and go. Do not embarrass yourself by pushing the matter to the eventual conclusion of a refusal. *Remember this:* it is difficult enough to launch a new project with an enthusiastic, intelligent group. With a lukewarm team behind you, failure is a certainty. *You* should have the strength to turn *them* down: you haven't lost a thing, so enjoy the luxury of saying, "Gentlemen, thank you for your time, but I know that this device could be of great value to some company, and with the mild interest you have shown, you could not handle it successfully. It takes great enthusiasm to launch a new product." Have faith in your invention and your ability to launch it properly. Keep trying till you find the company that does get enthused about what you have to offer. When this happens, you won't have to wonder whether it is the right one–you'll know.

How to Evaluate a Letter of Introduction

One of the most dangerous fallacies to be combatted is that "you can't get in without friends" or "they won't see you without a letter of introduction." A lifetime of experience will teach you this:

1. You cannot possibly know people in every company you need to visit.
2. Introductions are *rarely* to the right people in the company, and the wrong people actually impede your progress. *Only* by going cold can *you* choose.
3. Important business is conducted between strangers throughout every phase of industry—you do *not* need friends . . . you *make* friends out of strangers.

Remember: even if you do "get in" through an introduction, you stand or fall on your own presentation and its worth. Do it right!

Only if you are lucky enough to mingle with top management people can an introduction "carry weight," and the writer has found . . . to his amazement . . . frequently the "pressure from above" is *resented.*

Analyzing the Reasons for a Rejection or Mild Reception

There is much to be learned by each meeting, successful or not, and the results, comments, suggestions and criticisms should be reviewed repeatedly, as if you were panning for gold—for indeed you are!

The simile bears further examination, too, because, like the gold-panning method, most of the material in the pan is worthless, yet each meeting will have certain aspects which can be most valuable in reshaping the presentation methods, and maybe even the form of the invention, for the next try.

When certain knowledgeable people criticize your invention constructively, you may think well enough of their suggestions to correct or change in order to overcome all objections. Then upon resubmitting your device, with the critics' ideas worked into it, it may prove acceptable.

Remember, suggestions are not inventions. Pointing the way toward a more satisfactory device is undeniably helpful, but the man who literally invents the improvement in the physical sense is the inventor.

One cannot change one's invention to suit every prospect; therefore, one must change the prospect–a more practical and less costly procedure. Each successive meeting should make you a better salesman for your own devices. One particular element of selling which most inventors lack, but which the wiser ones try to acquire, is to learn to adopt the viewpoint of the prospect–the man in the business of selling such devices to the public.

It is only natural that the inventor has a viewpoint almost wholly selfish–he is concerned with *his* sale or license; *his* royalty; *his* problems–the prospect is not. Many a meeting ends in a stalemate because the inventor has not learned enough about what interests and concerns his prospect to speak to him in his own terms. Learning enough about your field to appear cognizant of what would interest your prospect would swing many a meeting your way, which might otherwise end unsatisfactorily.

The knowledge you will garner meeting difficult situations and handling them well is a form of education which, unfortunately, no university can offer. When you "graduate" into successful invention and exploitation, you may indeed be proud of your education, because this rare form of business know-how can place you in the class of the inventor-businessman, who earns $25,000 per year up to $100,000, and more. It is indeed worth the effort, time and persistence involved.

CHAPTER SIX

HOW TO TURN AN INVENTION INTO A BUSINESS

If you would learn the best way to start a plant of your own, study with the best teacher; our Creator. See him at work in your garden. Observe how the beautiful, mature plant of today, brings forth the seeds which become the plant of tomorrow. There is guidance for you, from the Master.

As the reader may have observed, turning an idea into a dollar by setting up a new business is a difficult job. However, for those who may be advantageously placed in business, full-time professionals or just people with a truly outstanding idea, it *is* both possible to do and practical. For others, it is doing it the hard way.

If that is the case, who, then, can succeed in marketing original ideas? The answer is: those who are *in* a business; those who find ideas a natural product born of the needs of the business or industry; people who have clear enough perception and long enough experience to *know* when an idea is needed, wanted, right in price and right in its physical form. Even for them it is not *easy;* however, it *is* possible! They *do* have the best chance of putting it over.

In Chapter Seven we study marketing your *inventive ability* in place of your inventions. Here, however, we shall confine our interests to manufacturing and marketing your own devices. When the invention and the inventor are of superior quality, this form of turning ideas into dollars is the best financially, but the situation must be analyzed very carefully before setting up such a new company.

How to Evaluate Your Patent Position

Because so much is at stake in starting a new company around an invention and a patent, we must make absolutely sure what our "rights" are. Let us therefore bring out some of the obvious errors which frequently occur, even among the more educated and experienced patentees. The drawings of the patent show clearly what we have invented, and to further clarify any possible misunderstanding of those drawings, the patent specifications describe both the physical form of the invention and its way of functioning. Does one have a right to assume that, having shown and described the invention, *that* is what the patent covers? Unfortunately . . . *no!*

The *only* part of the patent which tells you what you are protected on is *the claims* . . . and even these must be interpreted in the light of other patents. This sounds so completely unfair, and yet it is not. It is merely technical; the interpretation being necessary because so many inventors improve some other inventor's invention, and yet only the *improvement* is covered, not the whole. Thus, improving an invention allows you to stop the original inventor from using your improvement without paying you, and, conversely, it prevents you from using the original invention just because you worked out an improved version of it.

To illustrate briefly, your inventing a plastic attachment for a telephone receiver which secures it to your ear and shoulder to leave your hands free for writing, is a patentable invention. To describe the attachment alone would be an incomplete description; the receiver is part of the working combination. To describe the telephone *receiver* in the claim does *not* give you the right to manufacture telephone receivers. Many another invention has to recite, in all its claims, some elements which, while part of the invention, are actually *not* covered by the patent; you may point out the very words in your claims which say that it is covered—but it is not. *This is an important point to clarify* before starting a business around a patent.

If, in analyzing the combination of receiver and plastic hook, we find that we are free to manufacture our invention without either being prevented from using it by the phone company or any other patentee, then we are in a favorable position to start our business. If we have made but a slight improvement over a former, competing phone holder, this too must be analyzed.

In the telephone-hook case, the device could be made by an independent manufacturer without regard to the phone company. However, to examine a similar "improvement" invention with other complexities, let us consider this: you have invented a little colored stamp to be stuck to the back of a Polaroid film, so that instead of pulling it out, counting ten seconds and separating it, you look at a little blue stamp. When the blue turns to red, the picture is ready to separate from the paper.

Assuming that such a convenience was desirable, that color-changing stamp could not be used except with the permission and cooperation of the Polaroid Corporation, and therefore could not be established as an independent company. *With* their permission, you could set up to manufacture, but you would indeed be limited and dependent.

Laymen complain that they are told their patents are

dependent, yet nowhere in the patent does it say specifically that the claims depend upon any other patent. Don't let this fool you. Learn whether the patent is dependent, usually by so simple a means as studying the references used in allowing the patent (listed right on the patent, at the end) and also, by buying the file wrapper (for an attorney, only, to study), which is inexpensive. The ideal business venture is one in which the invention is made in a field where the inventor has long made his living, and which he understands thoroughly. In such a set-up, the invention has the added advantage of being marketed by someone whose know-how increases the worth of the invention. That same invention in the hands of some salesperson who must launch the invention *and* learn the business, simultaneously, would not prosper half so well.

The advantages of manufacturing and selling your own inventions lie in the simple truth that *you* understand your device best and sell it with a sincerity which buyers sense as being genuine—that has always been a powerful selling tool. Further, one good man devoting himself wholly to a product is better than 20 men who give it small, disinterested sales presentations. This is especially true when the item is just one of many that they are selling. Selling with judgment and enthusiasm is just as important an element of success as the invention itself.

From the viewpoint of profit, the man who goes into business to make and sell an invention does not operate as a royalty-paid inventor would, contenting himself with some 2%, 3% or 5% of the wholesale price. Instead, he has the scope of an inventor's royalty, a manufacturer's profit and a sales commission. The reason for the broader base of remuneration is that the financial risks are greatly increased; the money is not just pocketed, as an inventor would, but heavy business expenses must be met with that return. The inventor of a small electrical appliance may earn from $15,000 to $25,000 a year while his invention is in production; as inventor-businessman, producing his invention and marketing it, that same volume would pay him from $40,000 to $150,000.

All too often, inventors disdain business "talents" and over-rate inventive genius. A man named Maytag, inventor of the famous Maytag washing machine, once told a reporter that he created his machine because he saw a need for it. That one thing represented the only creative effort of his life. Despite his denial of being "creative," his creation of a giant industrial empire out of his one invention was a creative feat far in excess of his creation of the washing machine itself. Anyone who has tried to turn an invention into a business can well appreciate this statement.

Because Maytag was a pioneer in major electrical appliances, he had certain advantages which do not now exist. Still, we do have new appliances being introduced successfully today, in highly competitive markets. It is important to understand *why* the little fellows still can get a foothold despite the giants: The giants must have *volume.* G.E., Westinghouse, Sunbeam, etc. are at their best catering to mass markets. *Only* established inventions have known volume. In mass-production facilities, the giants have the little fellows completely outclassed. In establishing a new, unproven device, the little fellow can give invaluable personal sales effort.

The Casco steam iron was on the market long before G.E. came in with theirs. Casco was established by then. The new electric tooth brush, the Water Pik, portable hair dryers, electric can openers and countless other items proved that the small independent inventor is still very much alive and prospering. However, he is seldom an outsider without background in the field; he is not a man who walks off the street with a brilliant concept and walks out with a big check. The successful ones were *in* the business improving products; working in that field; developing features to use against ingenious competition.

Perhaps the greatest non-inventive benefit accruing to

a pioneer, is the building up of a *good trade name* which can exist, prosper and be sold, eventually, for a good price, even *after* the invention has long been made obsolete. The fact that the inventor and his new company *is* small is his greatest asset, for great companies cannot go in for the picayune profits of early missionary selling. The little fellow can!

If your "row to hoe" is a new appliance, and it is intended to establish a new area of usefulness for appliances, instead of just being an improved type of well-established appliance, then you may well question the rewards of doing all the costly missionary work and establishing a good business, eventually, only to have the industrial giants move in when it is at its best. However, if we study the constant, competitive process, we find that the little man has *a good, long period to enjoy,* and then, even *with* the competition, still prosper.

The extremely useful and well-merchandised Salton Hotrays have held their own against the giants; the makers of electric pencil sharpeners, electric staplers, can openers and battery-run clocks have prospered and grown in what we like to think of as "the American Tradition," despite the big corporations.

How to Set Up a New Business

Now, let's get down to the specific processes of handling a new invention, so that we may learn the many useful "tricks" and bits of know-how needed to establish a new business. Let us first examine the most advantageous ways of setting up. Please pay particular note to "which comes first; the hen or the egg" . . . sales or manufacturing? Our good sense tells us (in 95% of the cases) that *sales must come first. Only sales can make a business.*

Because many businesses are established with insufficient funds to set up both manufacturing and sales facilities simultaneously, it would benefit the business most to concentrate that limited capital on bringing in business, and using the already-established facilities of such machine shops, injection molders and other suppliers as may be appropriate, to make the merchandise for us on a contract basis.

It is a well-known fact that, with today's inflated prices for equipment and help, it is often better to buy than to make, especially when the new item has not as yet established a known, steady volume. A compromise could be arrived at by having all the component parts made on the outside and just set up for a relatively simple assembly and shipping operation.

As for sales, many a business has been started on a few hand-made samples and handsome color photographs of the items to show style, color and sales appeal. There are many things wrong with setting up a factory to manufacture a new item in the hope of getting orders and maintaining a crew to wait around for the eventual orders. When the investment in the sales force brings in the business to be manufactured, it also tells us *what* facilities we need.

Certain promotion-minded inventors have sold the idea of a neat arrangement for manufacturing to companies with the proper type of manufacturing facilities: they convince these companies—with the machines and the room to spare—to allow a new company to set up its offices in their building, and pass all orders through the manufacturing facilities at a preset level of costs.

Thus, the new company, while independent, as a company, gets to be a dependable selling organization to the older company, while the new outfit saves the set-up costs, factory leases and help, and concentrates its efforts on getting business. It really pays off for both companies. Many a company would like to have *two shifts* in a factory, with one set of overhead expenses. Here is the way to do it.

Another plan to start for less money is having the component parts made in quantities (which is cheaper) and stored that way, to be made up into finished devices when orders are to be filled. The number of companies both willing and anxious to compete for that kind of parts-making business is indeed impressive. No one cares who makes your devices as long as they function well. Certain types of inventions, chemical formulas and bottled goods can be placed with custom-packagers, who take your powder, fluid or pills and package it to order, pack and ship under your name.

For those who wish to organize a factory/sales company–a complete unit, ready to function as an established business–it is usually best to start with partners. A group can do better than any "one-man army," no matter how able he may be: a factory man, a salesman or manager, an inventor who can make, modify, change and otherwise keep things going and, finally, *when* possible, a financial man. Such a group has the makings of a business which will have staying power.

Many a new company is an outgrowth of some older, established firm, and the men may well have been the department heads of the former business–it had been a pattern for years and still is. Thus, they have plotted and planned their moves long in advance, and perhaps pooled their resources. Arranging for shares and division of ownership is a subject which it is important to get straight at this point: The author wishes to make clear, beyond all doubt, that the complexities and ramifications of business set-up and finance are far too important a matter to decide upon by taking a one- or two-chapter "course" in such a book as this. *Get professional counsel* from a good corporate attorney (not a criminal lawyer nor a divorce specialist), and, if possible, add an accountant to your team.

What You Must Pay for Money

Good accountants and attorneys can guide you, should you decide to raise start-up capital with a stock issue:

the field of money, investment banking, stock issues and private sale should not be guessed at nor approached casually. The usual stock issue for a new company, especially in these days of high interest rates is not the same as the boom-time, anything-goes kind of stock issue once possible to launch.

First, a big slice of the resultant cash is deducted for the firm or firms who launch the issue for you; frequently, in excess of 20%. Second, parting with common stock is parting with *ownership* of your venture. As many a company has learned these past ten years, corporations looking for acquisitions can buy up sufficient shares of your stock to gain control of your company, with or without your consent. Besides the advice your accountant and attorney see fit to offer you, *YOU must know* good advice when you hear it; you must accept the responsibility for your company. Study this phase of money-raising in the many textbooks available to you, so that you are not ignorant of the field when you are advised.

Borrowing from friends and relatives for business purposes has been done since time immemorial; borrowing from a bank is better, and relatives should be a last resort. Try not to borrow for a business wherein your major set-up costs are personal; should misfortune overtake your venture, it would be most difficult to repay a business loan as an individual. That is the primary reason for stock issue capital—your stockholders are taking a share of your risk in exchange for a share in your possible gains.

If you are one of the fortunate inventors who planned a device or piece of merchandise that lies within your financial capabilities to launch, you will soon learn that owning your own company is not some outmoded, old-fashioned idea. It is a great blessing, and a potent tranquilizer. Obviously, many inventions are too heavy and complex to finance personally. For these, the road to impersonal stock-issue finance is better than mortgaging personal belongings. If possible, conserve your money to live on during the normal, "dry" period of starting up. Even the best businesses require time to start functioning and bringing in the money.

While it is obvious that one cannot discuss "inventions" per se, and the setting up of a business for manufacturing an automatic telephone answerer is not to be compared to setting up to make a new type of ash receiver, it does become clear that *major* set-ups *do* justify stock; those set-ups which are within the capabilities of an individual to finance should be handled personally.

Because they are so prominent, certain investment bankers come to mind first: Lehman Brothers, Brown Brothers, Harriman, Goldman, Sachs, etc., yet these people are not interested in the small issues, unless they show unique promise. There are countless other investment bankers on various levels of importance who seek the smaller, starting companies and are properly set up to handle them. *Several* interviews with such potential sources of capital will give you a background from which to judge their appropriateness to your needs. *Always WRITE* for an appointment.

Such letters need only mention that you wish to discuss the financing of a new company with them, and it is most definitely recommended that the matter *not* be discussed in that letter. While it cannot be said that all bank managers are knowledgeable about guiding one to an investment banker, many are, and the houses you seek for the interviews will be known to your attorney, accountant and bank manager more readily than to you, for that is within the realm of their work. However, once you have located just one investment banker or even a reputable brokerage house, the names will flow–have the patience to "dig."

These are not the days for "shoestring financing." Even the term shoestring is no longer used. The best of inventions can fail in the marketplace if they are not given the advantages which money alone can procure–the sales efforts, advertising, proper quality of manufacture and, above all, the *TIME to succeed.* Even the best invention needs time, and can fail for lack of it.

It does become clear to any inventor of mature years, that he had better plan within the financial capabilities of his business life, if he expects to start his own firm. If he seeks partners, he is seeking for himself a smaller share of the ownership. This is still practical and viable. If he seeks a stock issue company, he had better study, long in advance, how such projects are handled, for he will find that his ability to handle the finances is more important, by far, than his invention, for without the finances, the project remains a dream, not a business.

One final word: learn what you need to from books by recognized authorities. Do not depend upon hearsay, casual friends' conversations and careless statements. Base your knowledge on a solid foundation–*read!*

CHAPTER SEVEN

HOW TO SELL YOUR INVENTIVE ABILITY TO AN ESTABLISHED FIRM

> **If a project becomes too difficult for you, do not consider abandoning your goal; instead, abandon the paths which lead to dead-ends. There is more than one road into Rome.**

THE complexities of selling creative work and patents has defeated many a stout-hearted inventor. Some were stopped by the very nature of the undertaking—they were equipped to experiment, not to sell. Others were stopped by the expense involved; many by their lack of social and business acumen in this most necessary function of the invention marketer (meeting top executives and presenting ideas effectively).

Does this lack of selling ability or financial know-how relegate a worthy inventor to the discard pile? It may, if he persists in launching projects beyond his selling abilities. What then *is* open to him? The answer is *bypass* selling, financing and all those duties which are so onerous to him and become an "insider," an employee hired and paid for *inventive ability.* From the inside, countless men of this type have made themselves highly paid, highly valued members of successful corporations. On the "inside," they found their lives both interesting and steady.

Selling Inventive *Ability* Rather Than an Invention

Selling inventive ability instead of an invention, gives a man greater scope, for no longer is his prospect judging one invention, but rather the ability of the man to work on many projects over the years and prove himself ingenious in developing *company projects.* They do not depend upon him to both create the project and work it out; usually the ideas come into the department and the inventive abilities of the man are used to work out mechanical problems, not marketing, selling nor industrial design problems.

Advantages of Becoming an Employee

The larger corporations of today have impressively large Research and Development Departments; many have large machine shops for developing devices to speed up production. R & D men can earn from about $15,000 to $25,000 per year, while supervisory department heads can earn some $25,000 more. In certain lines, Product Development is the most promising area for in-company efforts. Here the same range of salaries is paid, but the work requires a different form of talent, and college degrees carry less weight than business experience and background in the field.

College degrees still are a must, but ingenuity and experience in the particular area of research can be a major factor in hiring men. For those who do have the qualifications, the money is excellent.

Men who are good salesmen and creative too, do not always find themselves attracted to the uncertainty of being an entrepreneur and having the financial responsibility that goes with it. Such men, instead of selling their inventive ability alone, find companies which require the rare combination of creative ability and selling; this is the basis of many successes in merchandise lines, where creative salespeople, *seeing the needs* firsthand, create to fill those needs.

The pay of these creative salespeople is limited only by their ability to guide their employers into profitable fields. Many have earned $25,000 to $75,000 per year, and have even then left to become independent product design consultants. Here they *create to order* for their clients, and bypass the problems of first creating, then looking around for a buyer for their wares. These experienced innovators make a very good living; they do *not* amass sensational fortunes. Fortunes go with great risk. So do business failures.

Many of the successful product designers open an office, with their former employers being their first clients. The designers usually take on a staff, and the project becomes a team effort. This is good practice for product innovators with good ideas–only the man with a major, basic advance should accept the risk of a daring venture for big stakes.

One might ask, "Is an independent consultant or product development man not alone?" The answer is, actually, "No." He is part of a team–now retained per job rather than being paid a salary–but he operates as a single adjunct to a teamwork corporation, providing ideas and practical guidance, but still being part of a many-faceted organization comprising company facilities, personnel, patent counsel, artists and industrial designers . . . all awaiting the "spark" (and the sweat) that brings forth a new idea worthy of their money and efforts. From estimators to purchasing agents, it is amazing the number of people it takes to bring forth a new product.

One of the main attractions that the company-graduate inventor or idea man has for the corporate executive is that even top executives risk their positions when they launch products that fail, for the cost in time and money can be enormous. Knowing this, the executives seek people whose work has had a record of many successes, rather than accept a "good idea" from a source which leaves them taking the full responsibility for the new project. Corporate executives like to play safe.

As an employee, an inventor gradually absorbs *past* history as well as present status of the particular field in which he works. He advances with the field–he is not left outside looking in, with many false ideas of what may be needed.

Thus, the author is actually pointing the way to those who have been "outsiders" with imaginative ideas but little practical experience, as to how they may get the "inside" viewpoint and background from which successful inventions and ideas can be launched. Usually, the inventive type of person need only live in an atmosphere where industrial ingenuity is needed and rewarded to soon come up with ideas which are well suited to his particular employer. He sees genuine needs; sees places where improvements can be made and is *in* where there is always someone in authority to talk to.

For those who are already in such positions, and have ideas to sell or license, other chapters in this book have guidance to offer. With artful handling, the insider can normally find his way to the person of authority; however, the way to present the idea, the way to determine what kind of a deal may be open to a salaried employee and, finally, to handle the matter for greatest financial benefit, will be the subject matter of the subsequent pages of this chapter.

Being an "insider" automatically places the inventor in a position where he has assumed certain obligations because of the salary he has accepted. Before any discussion takes place with company officials concerning a new invention or idea, careful study should be given to the following pages.

Financial Arrangements for Employee-Inventors

When people are hired as employees to perform some technical or engineering duties, then standard procedure dictates that they be paid a salary, and for that agreed-upon sum, perform all duties normal to that occupation. However, "normal" is indeed an ambiguous word. What is normal for an engineer? If he gets an idea for a device which would be of value to his employer, does he "owe it" to him? Should he be paid or at least compensate for that new device beyond his salary?

Most employees do not expect to invent anything, hence make no arrangements for doing so. However, if you know that you *are* the creative type, and will strive to do some inventive or creative work for your employer, then it is advisable to make the proper arrangements . . . *first,* in an employment contract. If *no* arrangements are made, certain basic laws and customs govern your actions, and you may indeed feel cheated by them.

Obviously, if you were employed in a Product Development Department or a R & D (Research and Development Department) of a corporation, then whatever you invent during your business hours and with your employer's materials and facilities will become his, with no further payment due you. *If* other arrangements are stipulated *in writing, anything* both parties agree to may be considered legal. *In writing,* you can get a bonus for accepted ideas, a royalty paid on sales of your ideas or almost anything reasonable which you both can agree to. Cash payments can be made for assignment of patent rights; employment contracts can be offered as reward for certain outstanding achievements . . . all extraneous inducements designed to bring forth valuable contributions from talented employees.

Basically, the salary paid is to be commensurate with the level of the inventive man's ability; so, the need for other added inducements is made only to cover *exceptional* contributions, and not run-of-the-mill ingenuity which may well be expected of you.

The more elaborate arrangements are reserved especially for creative people who have proven skills and exceptional abilities. Few employers would offer those "extras" to regular employees.

One of the familiar obstacles to a trouble-free inventor-employer relationship has always been the *assumption* that an invention not in the employer's regular line of business cannot be claimed by him. If the inventor has used company time, materials or facilities to develop the extraneous device, he may find that the resultant invention may not interest his employer from the viewpoint of an article to make and sell, but the income from that device may, in the eyes of the law, be partially his.

When the individual inventor is a man of reputation or one with an idea worthy of major investment on the part of the employer, certain arrangements can be made which places the inventor in a position to benefit more substantially from his efforts, hence assuring the employer of that effort being expended. Such an inducement usually is in the form of a royalty.

Practical and successful arrangements have been made along the lines of establishing either a separate department to make and sell the new development, or, if very important, a new division of the company is formed to handle it. The inventor is given a drawing account against some guaranteed minimum royalty, and annual or semi-annual settlements are made. If so stipulated (especially at the beginning of a new project), the inventor is not expected to return or refund drawings in case of a loss.

How to Find Employment Opportunities in Development

While ordinary employment channels are too well known to warrant explanation in this book, mention should be made of exceptional areas of opportunity in engineering, inventing, designing and creating. The business section of the *New York Times* on Sunday has been known to have from ten to 20 pages of large display ads for all sorts of positions involving creative work.

The trade papers (magazines) have sections in the back with employment ads for the particular field in which they serve. In addition, there has sprung up, in these past few years, dozens of employment offices, each specializing in some technical trade; i.e., in packaging, several specialize in package designers, package construction engineers, managers of packaging for large corporations, etc.

Large corporations have their own personnel departments for recruiting help, and even when no help has been advertised for, openings can be made for the rare or exceptional man in development. Having a talk with one of the top people in such a department can guide you as to where the best chances for your talents may lie.

Many engineering graduates seek employment through the employment offices of their former colleges or universities, for companies still seek new recruits for their engineering and developmental departments at the source—the school. Courses and credits are being given at many colleges, in technical areas from packaging to industrial design, and those same schools place their graduates when possible.

How the Occasional Inventor Can Become Prolific

It has been a secret worry of many occasional inventors that they could not qualify as "prolific," nor could they bring forth ideas at will, upon demand.

A lifetime of firsthand experience in such matters has convinced the writer that *methods* of inventing do produce the necessary dependable, commercial-grade, creative work. The lone outside inventor lacks the stimulus provided by the *needs* of a big corporation; he trys to compensate for it by imagined needs and ingenious solutions. The big-corporation man is given direction, specifications and background, and therefore can do a dependable grade of work without waiting for some "flash of inspiration."

First, remove fear of failure from your mind, and ideas will come naturally with the stimulus the *needs* provide.

If you decide that you are one of the favored few who can make the grade as an employed creative man, then you should view your employer as your market and prepare to serve that market well. To do this intelligently, prepare yourself for "jobs unknown" by developing *sources* of useful information, inspiring ideas and, above all, *a knowledge of the past efforts* in your line.

Learn that the Patent Office was created to serve *you*... to give you—legally–free—the greatest inventions, and provide specific directions for their use... all available through patent copies. In this file of sources for many needed things, bits of knowledge, processes, addresses where the knowledge may be open for reference and other valuable things to make you a knowledgeable man or woman, all should be assembled, lovingly, at your leisure... *then,* you will approach a job with confidence.

Once convinced that *stimulus* begets inventions, you will see how some men can produce dependably over the years. Good ideas do not come out of a vacuum. Get where the action is, where there are *needs* to be satisfied–the ideas will follow naturally.

One final word: In seeking employment at interviews, ingenious, creative people are not supposed to walk in with an armful of patents to prove their ability. Keep uppermost in mind that you are selling your *ability to think,* not just some specific idea or patent. And, no employer expects you to have already developed a deep knowledge of his special needs, unless you present yourself as an expert in that line. The fields open to the "inside inventor" are indeed too diverse to prepare for specifically, but getting to know where knowledge can be found will serve all your purposes.

CHAPTER EIGHT

THE IDEA MARKETS: The Toy Field

Toy designers have a unique form of compensation: their products are purchased as an expression of love; their success can be measured in happiness.

THE toy field is a general term used to encompass all creating, manufacturing and marketing of playthings. It covers so broad an area that it would be quite impossible to examine it intelligently without dividing it and subdividing it into more manageable units. It would avail the reader little to consider the merits of checkers and bicycles, model planes and modeling clay or dolls and bubble-makers.

Analyzing Categories of Toys

To bring order out of chaos, we shall select areas of interest which hold promise for the man or woman seeking an outlet for creative effort. Had this entire volume been devoted to this one field, it could not have done it justice, so enormous is the toy field in its entirety. Our task will be to ferret out those categories of toys and those areas in each chosen category where creative efforts are actually sought and rewarded. The insight as to which areas, which companies and which form of toys can be created by outsiders, and sold or licensed, will be the subject to be explored in this chapter.

Perhaps the one simple truth which should be stated, despite its obvious value, is: those who desire to create toys or playthings for the field must confine their efforts to toys with *patentable features*. Many ingenious toys are created but are unprotectable. Toy *manufacturers* can produce such items and make money on them, but outsiders, inventors and designers cannot. No one, honest or not, will pay you for something which anyone can use free without your permission. *This is basic!*

We wish to select for our efforts lucrative specialties in which sufficient ingenuity is required to make the patentable protection a natural element of the design. Here we come upon the subtle differences between what is acceptable from "inside designers" and what is expected from "outside designers." Recognizing this natural division, common throughout the toy field, will save many an expensive, time-consuming disappointment for the creative toy designer.

Let us stake out the areas for the insiders and the outsiders first, so that we shall not be involved in projects which would not be acceptable. Even confining our efforts to the correct categories gives us many problems; however, being in the wrong categories wastes our efforts completely.

To demonstrate the insider vs. outsider designing, let us consider this actual case. Hasbro Toys brought out a tiny doll, about 4" high, dressed in custom-tailored, hand-sewn clothes; it was very cute, lovable and highly successful. It was know to the trade as Dolly Darling. Hasbro proved that there was a market for a tiny doll for little girls. It was original in concept and sold very well. Does this mean that you could design a doll—maybe smaller, cuter, with some other costumes—and sell or license it to some toy company? No! It does not. Had you been head of a toy manufacturing company or even a toy marketing company, you could have launched it and made money on it. As a designer, you are out–no patent protection.

Huge businesses have been established with *no* patent protection; i.e., teddy bears, pandas, etc., for anyone can *make* and *sell* toys with no patent protection–but, without patent protection, an inventor or designer cannot sell or license. To further clarify the state of the marketing of these so-called "unprotectable" items, let us examine the rest of the Hasbro set-up and why the Dolly Darlings went over so big.

The Elements of Success for the Unprotectables

Hasbro markets a *line* of toys. They *have* a large, well-established organization of sales representatives all over the country. Dolly Darling became just another of Hasbro's clever promotional items. They did not have to establish sales representation on the strength of having this "hot item." They fed it into well-established channels, and, as a good item, produced with skilled production know-how, it was hard to equal, no less copy, and the established channels got it to the market fast, first and with typical toy industry "hoopla" . . . TV advertising and trade papers publicity.

The success of this tiny doll made many a designer, and especially women, try to emulate or beat it. Unless they were insiders, they were wasting their time. Let us set up certain ideal "outsider's items" as examples of what would be best to develop in order to have the kind of toys which can be successfully sold or licensed by free-lance inventors.

Establishing a Yardstick for Commercial Acceptability

To distill from the many past successes a set of rules to guide us would soon bring out the fact that there *are* certain qualities the commercial successes had in common, and those characteristics were the very reason for their broad acceptance. First and foremost is that each was a distinctive feature toy. They were not just improvements on other toys or further adaptations of well-known devices—each was a toy which stood alone in its field . . . *distinctive and original in concept.* Such toys are both patentable and capable of license or sale.

Think of all the successes of the past, and how each can be called to mind for its distinctive qualities:

The Kiddie Kar	The Yo-Yo
The Slinky Spring	The Magnetic Whee-Lo
Lincoln Logs	G.I. Joe Doll for boys
The Barbie Doll	The Hula Hoop
Silly Putty	Tinker Toys

There were dozens of others, too, each one marketed successfully for a long time, although eventually giving way to a changing world with different interests.

Why would a large firm, such as Hasbro Toys, with a most able staff of creative designers, take a new toy on license from an outside designer? Only because that particular toy was so distinctive and so commercially desirable, that it paid them to do so! That explained their ingenious toy Sno Cone machine, which turns common ice cubes into delicious "ice cream" (actually shaved ice flavored with syrups). One could hardly confuse this toy with any other.

That same company brought out "Lite Brite," a black box pierced with spaced holes, and furnished with clear plastic pegs of many colors. By inserting the pegs into the holes in various designs, the light inside shone through the plastic pegs, forming a colorful, lighted, geometric design which was delightful to see. Despite the evident abilities of Hasbro's own staff designers, Lite Brite was worth taking from an outsider. It was ideal for TV and store window advertising. In other words, it was distinctive!

Inventing a Toy with Distinctive Characteristics

Both Lite Brite and Sno Cone were important enough to build business upon—they were *not just improvements on old ideas.* Hasbro and other major toy houses have paid out royalties to outside inventors which would impress almost anyone; many such checks ran from $25,000 to $50,000 a year for several years, and then continued at a lesser figure for many more years. As the basis for an independent business, Slinky Springs, Silly Putty, Whee-Lo, Yo-Yo and many others have paid off for seemingly endless periods, because those who sold them knew that while the toys grew old, the children were a new crop every year.

One old-timer should be cited as the ideal toy for both inventor and manufacturer—the Kiddie Kar! It was in the ideal price range; a really needed toy for the "romper set." Unlike remote-controlled cars and motor boats, it did not require a deep knowledge of engineering, but it did require *knowledge of children.* Women, particularly, are able to operate in this area and make it pay off. Inventors who operate in fields they know best succeed! If you possess this important knowledge of children, have the patience to add to it a knowledge of your markets and the selling methods taught in the earlier chapters.

Proof of the fact that knowledge of your chosen field is important is borne out in the marketing of the well-known toy, the Slinky Spring. When this flat, helical spring is worked by a demonstrator or a motorized display, people find it fascinating and buy it. Just lying in a box, it has little appeal except for those who have seen it demonstrated before. Today, salespeople cannot be depended upon for "selling" a toy.

Paid demonstrators *can* justify their expense in added

sales, and even motion displays can be made to demonstrate toys continuously for relatively modest money; nevertheless, today's markets have come to depend for their sales on toys which are obvious. The big discount toy stores, especially, need toys which can be piled up on the counter or floor and literally sell themselves. Knowing what is acceptable today is an important asset.

The very astute marketers of such hits as G.I. Joe and the Barbie Doll, deliberately set up million-dollar projects with costly tooling and facilities to keep out the little fellows. Such situations are for "insiders"—part of an able team—not a good bet for an outside inventor or designer. If you are convinced that you belong in this picture, seek a *job* in it. This is not a good area for an outsider to devote time to, unasked.

In using the examples of the two extremes, it can be seen that the smaller businessmen and the outside designers need each other to survive in this highly competitive market. An inventor with a distinctive, fascinating toy, protectable by patents, serves a more important purpose for the smaller firm than the larger one. If, in addition, it is relatively inexpensive to produce, that is a greater attraction to the small firm than the large one; as stated, the field's giants actually seek high-cost set-ups to take them out of competition with "every little Tom, Dick and Harry."

Within the toy and hobby fields, there has grown up a most unique area for ingenuity, but it is largely confined to engineers and the more technically trained designers: remote-controlled cars and planes; talking dolls and doll-voice mechanisms; gasoline-powered miniature engines for racing cars and boats. This is a most treacherous area for development work because, today, almost every manufacturer has turned to Japan or Hong Kong for both his designs and his manufacturing. This is a major outlet for inventions, but should be confined to those *in* the field who know their way around. America has become a *selling agent* for Oriental toys, and this has complicated the designer's selling enormously.

For those designers who wish to build toy businesses around their inventions, there is a modern trend which has come into being through necessity—the making of toys in the Orient, under contract, rather than setting up domestically for production. Almost all mass producers of toys today buy whole or part of their output from the Orient. American labor has outpriced itself in the toy market; oriental toys are delivered—with transportation and customs paid—for so much less than our own production costs, that we are not even in the running.

Through this change, the newcomer may be on a better footing with the older companies by having his toys made overseas. He buys *toys . . . not* equipment. He buys toys properly made . . . *not* the uncertain results of a new, untried manufacturing facility. He can concentrate his cash on selling . . . *not* manufacturing. This is a decided advantage.

How to Locate Firms

For those who are not already in the toy field, the author recommends that a study be made of the trade papers—not reading a single issue, but subscribing and absorbing . . . *over a long period of time* . . . all the necessary details of the field. The excellent magazine, *Playthings* (get the show number if you can), is published by Playthings Magazine, 51 Madison Avenue, New York City 10016. *Toy and Hobby World* is published at 1107 Broadway, New York City 10010. Through studying these two excellent trade papers, you can learn which companies make and sell the type of toys you have invented.

Two huge buildings form the toy trade center. One is at 200 Fifth Avenue on 23d Street; the other, right behind it, is at 1107 Broadway. Here can be found everything from chemistry sets to dolls, bicycles to checker sets, bubble-makers to party hats and all kindred lines, such as Christmas ornaments, Christmas trees, educational items for children, etc. You may *identify* the lines and firms you wish to approach for a job or a license at the toy buildings, but do *not* seek either employment or licenses from *sales* offices. Read Chapter Three for details for either type of contact with the toy companies–stick to the "rules."

Learn to differentiate between toys and games; between mere miniaturizations of well-known adult objects and "inventions." Do not confuse toy trains with toy planes, nor string-pulled toy autos for little children with motorized racing cars for young adults. Play is indeed serious to the older hobbiest. There are inside jobs galore in this field; buying and licensing of inventions and made-to-order developments for the established inventor. Yes, the toy field was chosen no. 1 for good reason; it is a huge industry built on ideas and ingenuity . . . an inventor's paradise.

For interested parties who wish to explore the advantages of having their toys made in the Orient, those very enterprising people have established the following facilities:

- Hong Kong Trade Development Council (Hong Kong Trade Inquiries)
  548 Fifth Avenue, New York City 10017
  Telephone: (212) 582-6610
- Japan Trade Center
  393 Fifth Avenue, New York City 10016
  Telephone: (212) LE2-7191
- Association of Tokyo Exporting Toy Manufacturers
  437 Fifth Avenue, New York City 10017
  Telephone: (212) 679-8970

American toy manufacturers have their own people go to Hong Kong and Tokyo regularly (usually the heads of the firms make this trip), and if a meeting could be arranged with one of these knowledgeable visitors, much could be learned, for the Orientals do have a most distinctive way of doing business, and such a conversation would indeed save false steps.

CHAPTER NINE

THE IDEA MARKETS:
The Advertising Display Field

The techniques of mass production have given America the highest standard of living in the history of man. Yet, mass production is possible only when teamed with mass marketing; for sales alone keep production going, and one of the vital tools of mass marketing is that tireless, silent salesman, the advertising display.

AN advertising display is a printed, mass-produced, sales-promotional device for selling a specific kind of merchandise; it is produced in substantial quantities and given to the retail stores which stock that merchandise. It may range from a simple, pictorial wall poster to a complex floor merchandiser; from a lighted and motorized window display to a display box in which the merchandise was shipped to the dealer; from a giant "spectacular" for a 1,000 pounds of cake flour to a tiny counter basket for 2¢ mints. They all have one thing in common: they are characterized by artistic and mechanical inventive ingenuity.

These display devices create for the national advertiser, a little "department" of his own, where his goods are given special attention. In the huge merchandise marts of today, displays are *the ONLY message the manufacturer can get to his customer,* for the store makes no attempt to conduct a study of advertised features.

All the displays are "invented to order," and the men who devise and design them must be able to "create to order" for any project laid before them. Many opportunities lie here for *employment* in creative capacities; as for inventing and then either licensing or selling the invention, that is confined to specialty items and devices with broad application.

There are several interrelated businesses, but they are far apart from the advertising display business described in the opening paragraphs of this chapter. There is, for example, the vast electric sign business, ranging from elaborate billboards and huge, spectacular signs down to the mass-produced, syndicated, motorized and lighted signs. Then the exhibit type of display is a field all by itself, also most ingenious in designs. However, *only* the advertising display which sells right at the point of purchase–the store–concerns us now.

The author, who helped create the floor stand business for the corrugated box industry, has seen this "depression baby" survive its humble beginning to become a huge and profitable branch of this industry . . . complete with its own sales force and creative staff. Here ingenuity is a marketable commodity *every day*–license and royalties are frequent; salary based on ability to invent and patent is more standard as procedure.

Advertising display stands must be shipped flat, set up quickly by laymen, and be beautiful to look at; the "cardboard engineers" who custom-invent such displays must be ingenious indeed, and frequently their work evolves into patentable inventions. If you find yourself attracted by this combination of art and clever mechanics, you may well find employment opportunities in the advertising display field. The so called "suppliers" who are the manufacturers of displays usually take on promising trainees and, with actual factory experience, train them.

As a thing apart, a parallel field exists in corrugated displays and also wire racks, and each has its own training of designers. It is considered most desirable to have graduated from a college course in industrial design or some branch of commercial art; training in engineering is most highly valued *if* the talent is combined with artistic sense. When you realize that creative people in the supplier's studios and laboratories are asked to design and custom-invent for such diverse needs as cigarettes, rubber balls, shot guns, flashlights, gelatin desserts and cosmetics, you can see that many diverse, interesting, problems are there awaiting imaginative artists and inventors.

A Never Ending Need for New Ideas

Perhaps the main attraction for the prolific creative mind lies in the never ending need for new ideas; if you are of the everyday creative type of mind, explore this unique field.

One of the advantages of this field lies in having customers who buy regularly–no "one shot" people here. Those who need displays to sell their products in January also require displays for February, July, October and December. Products must be sold . . . and that means new ideas, displays, ingenuity and creative business.

Over many years, brokers who merely created displays, then "farmed them out" to some manufacturing sources, were not in favor with the advertisers–they thought them parasites. However, with the constant broadening of the field, and the proliferation of newer materials, *only* the brokers could offer whatever was best for the advertiser, while the lithographer, corrugated house, etc. had to sell what they made . . . appropriate or not. These brokers soon had creative house groups of outstanding talents, and even employed outside specialists when the occasion demanded it.

The idea men and custom-inventors (called construction men) were employed at salaries of approximately $10,000 to start, and worked up, over the years, to between $15,000 and $20,000 annually. Those who turned their ideas and constructions into sales tools, and became sales representatives and brokers themselves, worked up to $25,000 to $50,000 a year. The independent outsider would get from some $500 to $2,000 per job for an outstanding (usually patented) construction. Had *he* sold it, that figure could have been doubled, but he would have taken a great *financial risk.*

Opportunities for Creative "Brokers"

Because the broker–the man who creates, sells and has manufacturing sources to serve him in *many* different fields–does not have to own a plant of his own, he is in the desirable position of being able to concentrate his efforts on ideas, inventions and the selling of them. It must be mentioned that he does have to *know his business,* for mistakes can be costly. But, he is in a most desirable position to launch his own ideas.

Most brokers started their careers in manufacturers' plants or studios; they learned their trade and made good for their employers before they started out by themselves. Thus it could be said that it was not much of a gamble to start their own firms. Frequently, they even had a "following" in the trade.

For the inventive-minded man, there are the display motor manufacturers, who not only make the fractional horsepower motors, but also devise ingenious gearing to get slow r.p.m. motions for the display designers, and, in addition, actually figure out the mechanisms (they call linkages) to get whatever tricky effects the advertiser needs, from his motors. This custom-inventing is most admirable, for it seems capable of almost anything needed. The manufacturers design gratis, and get paid out of the motor *order,* which can be quite substantial . . . motor, linkage and even the assembly.

Display brokers, who never use the term "broker," refer to their business as merely "the display business," and it is indeed broader in scope than such specialists as printer lithographers, corrugated box manufacturers or folding box manufacturers. Actually, the brokers become the business-getting agencies which create, sell and finance display orders for all of those prime suppliers today, and that is why they have special significance for the person entering the field.

If you find the field of advertising displays intriguing as a possible outlet for your talents, then you should first learn its "official" name–the point-of-purchase field–and then read its two best trade papers: *Spot* magazine, available from Schwartz Publications, Inc. of 6 West 57th St., New York City 10019 ($5 per year) and *Creative Signs and Displays,* published by Magazines/Creative, Inc., 37 West 39th St., New York City 10018, ($6 per year). Department store display is indeed different, for it is generally decorative and seasonal and does not sell specific products. Their trade journal is *Display World.*

Electric signs are a desirable field, too, but are not to be included in point-of-purchase displays, even though they are advertising specific products and frequently do have lighted and even moving advertising elements. It is the *large-run* display device which offers the greatest financial attraction for the man with an idea. It will confuse even the more experienced display designers to find so many specialty items being marketed in quantities and finding their way to savings banks, travel agents, business banks and others where *limited* quantities are sold, The kind of business wherein displays may be bought in runs of from 5,000 to 100,000 by a *single advertiser* to promote the sale of some *brand-name product*, is the field we must keep clearly in view. The small-run signs are sold by a completely different group.

An excellent trade association for the point-of-purchase field is The Point of Purchase Advertising Institute at 521 Fifth Avenue in New York City (10017). They publish a Directory which may be examined at their offices: here, you may see the entire list of members, what they manufacture or sell and the *many* forms of point-of-purchase materials.

Opportunities Among Display Customers

The list of display *customers* would be beyond belief: all the makers of brand-name products and consumer services, from railroads, airlines and steamship lines, to the advertising of foreign countries to attract tourists. Here indeed is a field where you can help the whole world sell its goods and services, with ideas and ingenuity.

To further clarify the employment opportunities, the prime suppliers—the manufacturers of printing or lithography; the corrugated box houses; the folding box houses; the die cutters (called mounters and finishers) who serve the printers and lithographers; the wire rack makers; the display fixture makers and the display motor makers—all need creative help and are in a position to train promising men.

The display brokers usually use men the factories have trained because they do not have the facilities to "factory train" potential employees. They *do* take occasional laymen to train, particularly if they show promise of one day being an asset in sales. The motor makers use men with mechanical rather than artistic backgrounds, and they require clever thinking and ingenuity to meet the great variety of problems in achieving "trick" mechanical motions to order.

Ads for help may be found in regular "help wanted"

columns, and also in the back pages of *Modern Packaging Magazine, Spot, Creative Signs and Displays, Display World* and *Package Engineering.* Also, there are many agencies in New York which specialize in all forms of creative help for the packaging and display industries.

CHAPTER TEN

THE IDEA MARKETS:
The Industrial Design Field

Industrial design is the wedding of beauty and utility, wherein the purpose is to bring forth products in which functional forms will have an inherent beauty, and which will need no extraneous, decorative elements laid upon them to be esthetically satisfying.

THE field of industrial design is indeed highly creative, but there are certain aspects of it which preclude its being treated at great length in a book whose purpose is to guide creative people in the marketing of their ideas.

Granted, this is an area ideal for making ideas pay, but *only* for those who have been adequately and thoroughly schooled for the task. Even though such men as Raymond Loewy, Henry Dreyfuss, Norman Bel Geddes and a mere handful of avant-garde thinkers brought the "field" out of the blue some 45 years ago—and at that time all one needed to get into this burgeoning field was talent and ideas—this is most definitely *not* the case today!

Specialized courses in industrial design have been

established in reputable schools from New York to Los Angeles, and even in high schools, universities, art schools and trade schools the subject is taught from its precollege aspects to its graduate, academically recognized industrial design phases . . . as a "must" for all aspiring, potential industrial designers.

Opportunities for Employment

It has even become a standard routine for the graduates of the industrial design schools to follow certain procedures in launching their careers; they all seek regular employment in one of the following diverse, yet closely related outlets for their talents:

- The studios of the established industrial design organizations.
- The industrial design departments of the large manufacturing corporations.
- Design groups specializing in label design, carton and package design, and trademarks, etc.
- Studios specializing in design of perfume bottles, liquor bottles, soft drink containers, spray cans, etc.
- The larger design groups handling major items such as trucks, tractors, power mowers, motorcycles, etc., and even hotel lobbies and exhibit halls.
- The professional model-makers who design for the industrial design organizations and never "go direct."

It becomes apparent, in this partial list, that the majority of the creative outlets are of an artistic rather than mechanical nature; the emphasis being on styling, good taste and ideas—not on scientific, mechanical or inventive lines. Industrial design is an excellent field for those who have been educated and trained for it. It is *not* recommended for those who desire to sell or license an isolated patented invention.

It is a custom in the industrial design organizations to employ various free-lance specialists for important jobs entailing work not ordinarily handled by the "house men." A very high rate of pay is offered to the specialists, but they are used only as long as the job lasts. An amazing variety of such specialized talents are available in the field: interior decorators, architects, exhibit designers, packaging construction men, clay and plastic model makers . . . and even "colorists."

Starting Your Own Firm

Despite the ability of certain industrial design students to finance an organization of their own after graduation, it is *not* advisable to start such a firm prior to having several years' experience in a well-established industrial design organization. While school does give one a basic knowledge of the art, only years of work in the field can develop the specialized knowledge of a specific business so that one can rightfully claim authority in it.

Inventors are frequently confused by the fact that industrial design firms *do* get patents. However, most of them are *design* patents, not mechanical patents. There is a clear differentiation between industrial design and invention, even though they both deal with ideas. This can be brought out clearly in this oversimplified example of an actual advance in the art.

When home washing machines first established themselves in American households, they proved most useful, but were indeed ugly and hard to keep clean. They looked like a piece of machinery, not a home appliance a housewife would take pride in possessing. Most had square or round galvanized iron tubs; crude, hand-cranked wringers over the tubs; water hoses dangling over the side and ugly black motors, exposed–truly a "workhorse" appliance, not a thing of beauty.

When an outstanding industrial designer conceived the idea of a smooth, decorative outer shell or "coverall," he did not affect the machine's washing functions, but he did indeed unify the mass of ungainly component parts and gave the washer a new appearance which housewives immediately accepted as far superior. That simple concept had what designers call "the simplicity of greatness" and the idea of the coverall found its way into every phase of designing, from typewriters, locomotives and motorcycles, to telephones, dictating machines and electric shavers. True, the "invention" had no moving parts, but it did serve a most useful purpose.

The keystone of the example of the way in which industrial design differs from invention is that the *industrial designer always starts with the inventor's invention.*

There are certain highly experienced industrial design organizations which are retained to create new lines for their clients. This is particularly true in housewares, office furniture and electrical appliances, where smart styling is more important than unique mechanical advances. These groups *never* get up ideas and try to find markets for them—they are *retained* to create them.

While *no* fortunes are made comparable to Ford, Maytag or Gillette in industrial designing, nevertheless substantial earnings have been, and are being made, by the top designers. The legendary fees of Raymond Loewy, Norman Bel Geddes and Henry Dreyfuss are presently a thing of the past, but *very* substantial fees *are paid* . . . and far more often than one might think. Perhaps the greatest payment to a good industrial design man is the exciting and absorbing creative work done for great corporate clients, and the resultant production in huge volume.

Career guidance for industrial design students is best found in good industrial design schools. There are many books, magazines and advertisements to spell out detailed information on remuneration. Here, in brief, it will be stated that salaries are from a low beginner at $6,500 to some $18,000 for a good veteran, then earnings of $15,000 to $50,000 for an industrial design studio owner or principal.

When industrial design men get into interior designing, their excellent fees must be analyzed as a separate and distinct field, apart from traditional industrial designing.

CHAPTER ELEVEN

THE IDEA MARKETS:
The Greeting Card Field

Never underestimate the therapeutic value of regular employment and its implied acceptance. Creative people *need* this life-enriching form of compensation even more than money. Acceptance produces artistic growth—money follows naturally!

PERHAPS nowhere else in industry is there so enormous and endless a demand for ideas and inventions as in the field of greeting cards. However, seeing the field from the viewpoint of an "insider," places the opportunities in a position to be evaluated by those who would like to benefit from the creative field which has built greeting cards into a multibillion-card annual business.

So many outsiders send in unrequested novel ideas, ingenious constructions for cards and poems and sentiments to such leaders of the field as Hallmark Cards, Inc., that there is an obvious need for authoritative guidance. Today, most of such efforts are wasted.

How Needs for the Types of Cards are Determined

To place the industry's needs and facilities in proper perspective, it must be stated that the lines of cards are scientifically created to meet the needs of the buying public. Market Research Departments at Hallmark, as an example, determine from their study of sales, just how many categories of cards are needed for each greeting card occasion, and precisely which people they must be designed for; i.e., in birthday cards, there is a need for cards for mother, father, sister, brother, aunt, sweetheart, employer, teacher, etc., and each category breaks down into age groups, too. Obviously, a card to a six-year-old brother bears little resemblance to a card for a 60-year-old brother. Research even guides them, accurately, to estimate the number of cards to be produced in each area.

Top executives, known as line planners, work with their own house-employed art directors, visualizers and supervising copywriters to lay out the jobs to be done. In a business running into the tens of millions of dollars annually, they could not depend upon chance contributions from outsiders. *All the work is created to order, for some specific purpose;* the sentiment is dictated, the character of the art is dictated, even the price is set in advance. Creative people *do* have a chance, but not by haphazard submissions.

The regular Hallmark staff is comprised of several hundred artists, cartoonists, letterers, art directors, visualizers, writers, poets, gag writers and inventors of animated greeting card novelties.

New Ideas for Greeting Cards

This "stable of talent" creates solely for the needs of Hallmark, making precisely what the line planners ordered. Despite these ample facilities, outside talent *is* used, but the card company commissions the work to be done exactly as they wish it to be. A marketable creation in this field is *not* just some clever idea received in the mail, unrequested, and placed in the line. The rare contributor who does have an idea worthy of finding its way into the line may get a check for somewhere between $25 and $100—hardly a goal worth striving for.

In contrast, many outsiders are used for work which the staff cannot provide, and these people, working on requests, are indeed well paid. However, they are almost always celebrities; i.e., Churchill's paintings, Eisenhower's paintings, the Charlie Brown cartoons, Walt Disney characters, natural color photos of great cathedrals, Bannister's comic photos of babies and Norman Vincent Peale's epigrams.

Upon examining greeting cards as a potential *field of employment* for people with ideas and talents, it is indeed ideal. The inventive man who can design practical, animated greeting card novelties, can, with no doubt, find a constant demand for his creations. The gag writers and cartoonists find great satisfaction as members of a talented team. Certain independent studios offer employment and free-lance work, catering to the small manufacturers, for they buy their line cards from these studios, even though they too, specify what they want.

New Ways of Printing or Producing Cards

The areas of the greeting card field which seem to offer the greatest opportunities for the mechanical, patentable forms of invention, do not lie in the creating of the cards, but rather in their production. Here are inventions which must be sold or licensed, and which produce the major amounts of money in which our "man with an idea" is primarily concerned.

It should be pointed out that those inventions which

find a market in greeting card production, also find a market in other fields of printed matter, such as the advertising field, direct mail field, display field, and commercial printing. The greeting card field, however, is the most constant seeker of the new, the beautiful, the unique, and that makes it most desirable for the inventor of a patentable form of printing. Further, the inventions in new printed effects have far longer lives and greater earning power than the usual "trick construction" of a card.

To illustrate the kind of inventions which have found success with the greeting card field, let us take the example of a printing process developed by an "outsider," and brought to a practical and useful stage of production under the sponsorship of Hallmark. This process is a form of printing which, when subjected to heat, causes certain of the printing inks to swell slightly, thereby giving the appearance of high embossing, without the need for costly dies. Although the concept and development were the work of an outside inventor, the project was sponsored and assisted by Hallmark, and their staff artists and technicians gave the idea far broader scope than any lone inventor could have afforded to give it.

Another basic patented process, this one invented by the author, was a new form of printing which made possible two separate technical advances: 1. Printing on one side of a sheet of tissue, and having the color on both sides. 2. Printing continuous stripes of color, making them into multilayered blankets as used in the well-known "honeycomb paper" novelties long used for party goods. This continuous-stripe process, when the honeycomb paper was expanded, became multi-colored figures, such as Santa Claus or clowns, or other useful and colorful room and table decorations. The author brought the concept and some crude, handmade models to Hallmark, and was given a development budget to explore it. The work paid off in royalty checks of many thousands of dollars over the years. Hallmark financed this entire heavy project.

One of the greatest "success stories" in greeting cards, is the 3-D picture using a "lenticular lens" pressed into plastic sheeting. It was used only for small displays and political election buttons for years before it managed to find its way into high-priced greeting cards—patience and endless development work was necessary to accomplish it. This "overnight" success took years to put over. The Vari-View manufacturers deserve credit for much of it.

The one thing these three unrelated processes have in common is that each produced a distinctive, basic advance . . . patentable and obviously salable. *Many* other such printing processes, embossing processes and decorative appliques have gotten their respective share of the huge greeting card market and held it for years. The procedure to be followed for the market of processes, related machines and other means of obtaining unique effects, is adequately covered in Chapter Three, for the fact that the items are different does not change the selling methods.

How and Where to Find Employment

As for the men and women who wish to seek employment in the creative departments of the large greeting card companies, do *not* send your artwork, novel constructions or poems to any prospective employer—rather, look up the correct names, addresses and personnel in the company you wish to approach, and *write* to them asking for guidance in applying for whatever category of position you seek. *When* and *if* they request samples of your work, try to *go there* to present them; it is indeed rare that people are hired by mail.

While there are innumerable greeting card companies, the most creative ones are best for employment opportunities. Look up the details necessary for writing in either *Standard and Poor's Directory* or *Dun and Bradstreet*. Get correct names, addresses and, when possible, persons to write to. Here is a list of the top half-dozen firms:

Hallmark Cards, Inc.
American Greetings Corporation
Norcross, Inc.
Gibson Greeting Cards, Inc.
Rust Craft Greeting Cards
Buzza-Cardozo Greeting Cards

Always remember that firms hire *people* as well as artists, inventors and writers; when you can, *go* for a job. Never just wander in; write for an appointment to make sure you are seeing the person with the authority to hire you.

For additional clarity, on page 129 reference is made to the artwork, poems, epigrams, etc. of *famous* people whose names are used to create additional sales. Just submitting better artwork or poems would not be acceptable to the houses publishing those cards. If you can do better work, seek *employment*.

CHAPTER TWELVE

THE IDEA MARKETS: The Gadget Field

Gadgets are the little conveniences of life which add so much to our comforts and so little to our expenses. It seems unfortunate that no one ever became famous for inventing gadgets.

THE term "gadget" is used to cover any small, ingenious device which is not of major importance–a solid-state transistor is small, but not a gadget; a tricky bottle opener or pipe cleaner is a typical "gadget." Despite this term of derision, gadgets that are of sufficient ingenuity *are* patentable, and many a fine gadget has earned more for its inventor than had it been a complex, hard-to-finance machine for more limited markets.

The Random House Dictionary lists the terms "gadget" and "gadgeteer" (inventor of gadgets), but nowhere can you find the proper appreciation for his efforts, except, perhaps, here. A good gadget can be launched for a modest investment; it can be sold with a simple display (because it is obvious in its usage) and, finally, it is relatively inexpensive to patent. Gadgets *are* important in business! They are sold in dozens of retail outlets, from drug chains and 5- & 10-cent stores to hardware, stationers and department stores. *Everyone* needs many gadgets every day.

When all wine bottles used corks, there were more devices for pulling out corks than you would believe, *unless you searched the patent files, where you would have uncovered them by the dozen.* Later, when metal caps replaced many of the corks, the gadgeteers were off and running again, in a new direction, made necessary by the change. Still later, canned beverages needed piercing devices in order to pour, and inventors worked out dozens of them. Then came the pull-tab cans and new gadgets fell sharply. Still, the gadgeteer had some 50 years of prosperity first. The pull-tabs were no "gadget"–they were a major development by giant corporations.

In contrast to the single, widely marketed gadgets, such as the opening devices, there is the line of products with a specialized market–broad, but not universal. A most ingenious man named Dritz, specialized in what department stores called "notions," a whole line of sewing and handwork gadgets which met the needs of housewives and seamstresses. He marketed his gadgets, each on a separate, illustrated card, making their purpose plain in pictures, not just words. By making and selling his devices, he earned many, many times as much as if he had licensed them for others to make and sell, and he kept the devices producing long after a less-interested licensee would.

Today, especially, the stores do not want small, cheap merchandise to sell; they want "big ticket" items–even the old five-and-tens now sell things up to $50 a unit, and more: office files, chests, organs, kitchen cabinets . . . all expensive items, but not good, old standbys like single pencils (they like to get pencils in 20- and 30-pencil packs). So, the "penny gadgets" have become passé, and the ones producing greater sales and greater profits are now "in."

The Best Way to Make Money Out of Gadgets

While obviously, there can be no blanket rule covering gadgets ranging from automobile ash trays to tricky telephone-number index pads, it is relatively safe to say that to make and sell your own devices would be to give them the best chances of success. If starting a business is not practical, your next best procedure would be to sell to or license a company (*not* necessarily a manufacturer) now successfully marketing a line of related products. Some astute organizations manage to do a big business with very few people, by selling to the big chain stores where one man can cover them all. Here too, such firms can sell "special deals" to supermarket rack jobbers who handle nonfood items in their own racks (hence the name) and like to feature obvious, useful gadgets.

A midwestern manufacturer, Ekco Products, mentioned before, handles an impressive-sized line of kitchen gadgets, each mounted on a colorful, "in use" card which shows the gadget in an unmistakable manner. Many of Ekco's items sell in the millions, and go on, year after year, which is easy to explain because they still are the best devices on the market for their particular purposes. As an example, Ekco Products markets a long, narrow paring knife mounted on a swivel. This narrow, curved blade has a long slot in it and the slot is the paring edge. Housewives have bought these paring knives by the millions each year, and still continue to do so. That is why a good gadget is a joy to make and sell–a minimum of trouble for a maximum of profit.

Introducing a "Line"

While the example of Ekco Products with a large line of kitchen gadgets seems to infer that a *line* of products is essential, the author wishes to state that lines are built up over a period of time; it is not safe to bring out a whole line of untried products at the beginning. Rather, bring out *good* items, slowly, and market-test them. Good lines are evolved naturally, not created in one idea jam session. There are obvious advantages to working in one type of retail outlet, especially from the point of view of selling and display.

The governing factor for gadgets is their retail price, and this usually dictates how they should be sold. Many an automobile gadget warrants a sales force for one item or a small line. Other types of items, such as garden gadgets, gift-item gadgets and sell-on-sight novelties, have justified high sales expense by becoming a vogue . . . for a limited but profitable period of time.

One of the good places to launch single gadgets is in the field of specialty advertising. Here, useful, simple gadgets are produced in large runs and are sold in small quantities, imprinted with the advertiser's name and copy. Quantities are impressive and patent-life long. The late Charles Ward of Brown and Bigelow, St. Paul Minnesota, invented a desk calendar device over 30 years ago, and it is still "new" and popular today. Millions were sold, and that, plus his other ingenious advertising ideas, made him a millionaire. Osborn, Kemper, Thomas of Cincinnati, Ohio are also purveyors of ingenuity in the remembrance advertising field.

One of the truly outstanding men in this field is a Chicagoan named Marvin L. Rubin. One of his companies, Sparky of Chicago, makes, among other things, advertising ties and advertising aprons, which are sold to big corporations—gas station attendants wear ties with the company trademark worked into the design; chain restaurant and roadside restaurant waitresses wear aprons with the company name artfully embroidered upon them. By making and selling his own inventions and ideas, he has been able to earn many, many times what he would have earned as a 5% royalty inventor.

Many injection molders seek items which can be sold in chain stores as a staple item. Others offer their wares to the specialty advertising trade. The stationary, hardware and notion departments are ideal for gadgeteers. Look for these attractive counters, which have come to be housed in the chain store basements. There–among housewares–are gadgets for curtains, drapes and shower curtains; closet door attachments for hanging ties, belts, shoes and trousers; trick hangers for women's garments with shoulder straps, special men's trouser hangers and other specialized hanging devices; picture-hook devices for all manners and kinds of pictures and all sorts of new gadgets for applying paint to walls–a gadgeteer's paradise!

Once again the potential gadgeteer must be cautioned that a first device must have *exceptional* merit to warrant starting a business around it. When the eventual line becomes a reality, a number can be added to enhance the line, which may not be such a "world beater" but will justify a *place* in the line. For a beginning, the idea should not only be ingenious and original, but also must serve a genuine need which could develop into a huge volume. You may not capture the huge market, but if it is there, potentially, even a small part of it will be worth fighting for.

How to Make Sure Your Idea Is New

As recommended earlier in this book, a *personal* search of the Patent Office will not only provide a clear picture of the state of the prior art, but also will *be an educational experience* worth far more than you will have to pay for it. In the *examiners' file,* which you may search, too, the examiners have not only assembled the domestic files, which you can examine in the main search room, but also they have added foreign patents pertinent to the art . . . also, catalogues from commercial producers, and even articles from trade papers and ads clipped out and mounted. Thus, certain things which had been used, but not patented, are known to the astute examiner, and should be known to you. The more you learn, the better able you will be to develop gadgets which will succeed commercially.

Try to confine your efforts in gadgetry to fields in which you have firsthand knowledge of the public's needs. There are so many areas which invite your ingenuity that there must be one, at least, in which you would prosper. Here are some of the fields in which gadgets are needed, and some successes to guide you.

- Specialty advertising devices for office and home use.
- Automobile accessories and conveniences.
- Kitchen gadgets for food preparation and refrigeration.
- Gardening gadgets for home gardens and city terraces.
- Electrical gadgets for wiring, plugs and lamps.
- Photography gadgets for albums, frame mountings, etc.

Where to Find Opportunities

There are several invaluable areas of reference for the field of gadgets, and it would be a mistake not to learn about them, because so many gadgets and novelties are sold by mail through ads in the women's publications, that to miss them may cause you to invent things already on the market, but not patented.

The Miles Kimball mail-order catalogue is full of such material; send to them in Oshkosh, Wisconsin for a free copy. The Sears catalogue, while having fewer gadgets, does have a wide range of interests. In the automotive field, there are well over a dozen magazines for the men who make motoring a hobby . . . just full of gadgets–all worthy of careful study.

In the Main Library of the New York Public Library System at 42nd Street and Fifth Avenue, there is a Magazine Room with *all* trade magazines available for reading *there.* In almost every large city in our country, similar trade paper reading rooms are available. Avoid the one-shot cramming session. Look over the magazines to find the ones which help you the most and subscribe to them.

There is no excuse to be ignorant of unpatented ideas or devices whose periods of protection have expired. It is far too costly to invent things which had existed before (and sometimes had been very successful). Library reading rooms are free, and endowed expressly to help you in your life and work. Use them!

A unique mail-order section for gadgets and novel merchandise has developed in the rear pages of the magazine section of the Sunday *New York Times.* Also, in the high-class men's and women's magazines the advertising of mail-order items has become a truly enormous, impressive market. Study these areas for clues as to what sells and what type of appeal you need to make people bypass a large and convenient retail marketplace to buy products by mail order. They *must* have something which the regular stores do not offer. What is it?

Some magazines cater solely to canvassers and door-to-door salesmen. They sell gadgets and eye-appeal products, often being required to purchase the items they sell, just like a storekeeper would. Sometimes they get a refund of "sample money" after orders come in to the manufacturer.

For high-class, costly gadgets, New York City has the beautiful shop of Hammacher & Schlemmer on 57th Street. Their display of ingenuity is indeed inspiring to an inventor or gadgeteer. Shopping is part of a gadgeteer's education—don't miss it!

Remember, it never pays to copy! See the *need*—not just the gadget which catered to it—then create an even better gadget of your own.

CHAPTER THIRTEEN

THE IDEA MARKETS: The Packaging and Packaging Machinery Fields

Nowhere in our American economy has the growth of any one field so affected our everyday living as in the field of packaging. Parenthetically, nowhere in our economy has an inventive concept found acceptance into so many lines of business, for it offered the basic advantage of being able to place manufactured goods in a safe environment until their eventual distribution, sale and use. Thus, packaging was the keystone of mass production, and mass production is the keystone of our prosperity.

Opportunities in Packaging

As a broad, basic concept, packaging is the placing of merchandise into a container for the purpose of protection, convenience and handling, in its passage from maker or packer to user or consumer. In its specific applications, it varies widely with the merchandise it was designed for. When one says packaging, he could mean: folding cartons, corrugated boxes, cans, collapsible tubes, glass bottles or jars, plastic or paper bags, wooden boxes, fibre cans or shrink-film packs. With each category an industry in itself, it is indeed apparent that packaging, in its broad sense, has become a giant field, and in its growth, has created opportunities for endless numbers of talented and imaginative people.

Finding New Ways to Mechanize Packaging

If one can lay stress upon the one characteristic most attractive to inventors and idea men, it is that packaging has become a rich source of income for those who have mechanized their packing procedures. The enforced education of American youth, combined with the desire for big-pay jobs, has left a severe shortage of unskilled help, or even semiskilled help, for packing by hand. Competitive products have forced manufacturers to reach out for talented people to find ways of mechanizing their packing lines . . . even to products once thought too irregular or unhandleable for mechanization.

The writer sincerely believes that, with the patience to divide and subdivide, study and appraise, areas appropriate to your efforts can be selected (preferably, near at hand) for developing packaging innovations worthy of your investment of time and money. Being so broad a field, one can select items within the mental and financial capabilities of the innovator. At this point the term packaging has been used as a "coverall," but we shall now start to divide it so that we can analyze its opportunities for the readers.

The Package–The Packing Machine–The Packaging System

Let us start by dividing the package from the machine

which packs it. The package may be, for this example, a simple folding carton to be used for granular soap. It may have a patented, rip-open area which turns into a pouring spout. It is paperboard and printed. The erecting machine for the carton takes a whole stack of the cartons, in flat, tubular form (each made as a tube pasted to itself), expands the tubes into square form, one at a time, and passes them over glue wheels which spread glue over their bottom flaps. Then it presses the flaps together to form the bottom of the box. Now we have an open-top box.

The soap powder box-filler, may receive its endless supply of powder from the floor above, down a chute. As each box passes in front of the machine, it triggers a device which pours a measured amount of granular soap into the box.

Finally, the loaded, still-open box, passes through a machine which spreads out its top flaps, spreads glue on them and seals the box closed.

From there it goes on to still another device, for packing the closed, filled boxes into corrugated boxes.

Analyzing the Packaging Process

Now, let us analyze what we have just "seen." Yes! The box *was* simple. True, it had a very ingenious spout to pour the soap from, but it was really a plain, rectangular box until the housewife opened it and made the spout herself. Yet, for a cheap, high-volume item like soap granules, that carton would be worthless without machines to load it inexpensively and turn the flattened boxes into set-up, ready-to-fill boxes.

When a box and machine combination is necessary, it is called a *packaging system*—no part can function alone.

This system incorporates the *work of many, different inventors:* the feeding out of the single blanks from the "hopper" (stacks) was one man's work . . . the gluing mechanisms were another's . . . the measuring of granules still another's . . . the various complex feeding and timing mechanisms, probably the work of several more . . . and the sealing, again, another's. Such systems *evolve* over a period of many years. Inventors make their excellent contributions in box and machine, but in time, each is obsoleted by improvement, but many pass into the public domain to be incorporated into the later, improved version of the device.

With all its complexity, it soon dawns on us that *nothing* can be done until the box designer comes up with a package the customer accepts. He does *not* do it all–he designs the box and spout, but without it *no designing* of machines can proceed.

Frequently, inventors become discouraged by large, complex developments, not realizing that they have evolved over a period of time from the work of many contributors. One of the good training grounds for folding box designers (construction men) is Container Corporation of America. Here, one of their top men assembled several volumes of actual folding cartons, each pasted to a page, with an explanation beside it. This provided countless pieces of ingenious mechanics to meet the needs of many customers.

Why then, does this encyclopedia not turn all men who follow into mere copiers? Because learning all the basic "tools to work with" does not preclude ingenious use of them. Each problem . . . each article to be packaged . . . each package to be mechanized, requires its own solution. As in machinery, having many forms of gears, shafts, motors, cams and transmissions does not stop an engineer from inventing a machine with an original purpose.

The Book Shipping Box Success Story

Recently, a talented designer invented an ingenious shipping box for books. Instead of trying to get a royalty for his reward from a box maker (the usual procedure), he made a connection with such a house as *his* supplier, and *he sold the boxes* to publishers instead of licensing them, assuring himself of dependable sales efforts, as well as much higher profits. With contacts in the publishing field, he soon learned of their packing problems—an advantage provided by close association with his customer. As an inventor working through a supplier's salesman, he would not have learned the vital facts which only direct contact can provide. He learned that hand-packing was rapidly becoming far too expensive, and publishers were looking for a way to reduce costs.

This inventor, working alone, conceived a way to machine-pack books in his own corrugated shipping folder, and in his own apartment, built a wooden mock-up of his proposed packer, complete with hand-worked, moving parts. With this crude but obviously clever way to handle books safely, he managed to sell one of his more important publishers on backing him financially, so that a regular steel prototype could be made by a good machine builder. After much hard work in cooperation with this machine builder, the inventor finally had a perfect working model to install in his publisher's plant. The first machine, despite some faults, was a good, working production device and was accepted.

Since that first machine, the inventor has made many more, and each one was improved over the last. Under the name of Libra-Packer, the inventor, Hans L. Levi, gradually sold America's leading book publishers on his machine packing device—a long difficult, step-by-step effort which turned a dream into a solid reality. This story should encourage others whose abilities need a profitable outlet; complex undertakings *can* be handled step-by-step. Hans Levi, a German-born, naturalized American, fought against the Nazis for his adopted country and built a life for himself and his family with ingenuity and fortitude.

As an important "aside," the writer, a far more experienced package designer, was retained by a large folding carton maker to develop just such a book-shipper. The development was equally good, but no one at the carton company was big enough to do what Mr. Levi did. First, by *selling his BOOK SHIPPERS* (with a royalty and selling commission in the price) instead of selling a *LICENSE for someone else to do it for him,* Mr. Levi was "in" and established long before the writer could close a license deal.

In sharp contrast to this "heavy" type of development, designing packages which are of simpler requirements *far* outnumber the heavy ones. For men and women who want to learn a business, packaging is full of opportunities. Besides folding cartons, the field now offers careers in Styrofoam containers, injection-molded containers, skin pack, shrink wrap and countless other branches of modern packaging. There are countless outlets for design, invention and idea creators. Each, however, is a specialty needing study and experience of a specialized nature, learned at some manufacturer's plant. Such fields have great possibilities for discovery and invention: first, because any new field has, but second, because when the "gold mines are being staked out," *that* is the time to be there.

Another new area just opening up is the molding of various foam plastics, by heat and pressure, into handsome, useful forms. In one case, the material, like a soft blanket, comes by the roll, and is molded by male and female dies and heat into such items as egg cartons, hot-food delivery packages, etc.—and here, even the hinges of the covers are molded right into the structure, then nested for shipment, with not a hand necessary for finishing the job. While the *process* is already patented, the items which can be designed and patented are endless.

The Egg Carton Success Story

Perhaps no other folding carton is as familiar as the egg carton. For well over 50 years these useful containers have taken eggs to market and to the home, until we all have been exposed to their ingenious designs. There has been a never ending design-structure evolution in this burgeoning field, and countless egg cartons made their successes, then gave way to others. One, the "Egg Safety Carton," achieved a very long life. It was a most ingenious piece of mechanics, folded from a single blank of board. However, with all its ingenuity, it was the fact that a machine for erecting it quickly had been created which accounted for its then unprecedented popularity.

In the 1920's, a very perceptive inventor named Burger realized the necessity of getting away from the slow and costly hand-erecting, and developed a simple, hand-operated device which, with a single pull of its lever, set up the carton from a flat, pasted blank. By equally clever patenting, he came to control the field for some time, and his astute mind brought millions and millions of egg safety cartons on the market before competitors, who were caught napping, could regain their trade. Finding such vital needs and filling them *can* produce big money.

It was the desperate efforts to get out of this "bind" which brought many other talented inventors into the fray and opened the field along newer lines. Mr. Burger had a *long* period of control, and well-deserved it. Despite temporary financial discomfort to his competitors, he did benefit the field by giving them a quick-to-erect box; otherwise, perhaps some molded product might have scooped the whole field even then.

After 50 years of development and sales, the paperboard egg carton is still being sold, but inroads have been made by molded-pulp egg trays, Styrofoam-molded trays and vacuum-formed, clear plastic (in an effort to let the buyer visually examine the eggs for possible breakage). As usual, heavy competition spurs invention.

The Can Band Success Story

Although the egg carton was a form of multipack (several like items sold as a unit), when the soft-drink and beer people started their "six packs" it was considered an original idea, for the carriers and the loading had no relationship to the egg box techniques. Atlanta Carton Co. built up a considerable business in this field, and while there were others, and even earlier ones on the market, Atlanta (later called Mead Paper Company) had excellent mechanical ways of prefolding the complex blanks and gluing them. Developing mechanical set-ups insured their success.

Added to this, machines were developed, later, after years of manual setting up of the carriers, for a quick, automatic erecting of the flat-shipped carriers . . . ready for loading on the quick-moving production line. Without going into detail, the writer had some of the best and earliest of the six packs, but failed to find backers to finance the mechanical setting up and score-breaking; hence, he was unable to make a financial success of the early start. Later, with more years of experience, he did develop a most subtle carrier which became the standard of the industry, and this was handled better . . . thanks to previous failures and the lessons they taught.

At that time, breweries were being forced to speed up their bottling and canning lines, but the complex operations of loading the can carriers made 350 cans a minute maximum production. The locking tabs of that carrier were the obstacle, and this created a stalemate which, had it not been overcome, could have caused the brewers to switch to some other form of carrier. The writer, who had been developing other unrelated devices for Container Corporation of America, saw this plight and invented a very simple but extremely tricky "swinging door" lock which the incoming can pushed open . . then the lock closed by itself after the can passed.

This critical bit of design made it possible to step up carrier loading from the previous maximum of 350 cans a minute to 650. With the help of some able patent attorneys and many dollars, two U.S. patents were granted, and the invention sold to C.C.A. for well over $40,000.

Much later, The Mead Corporation brought out the concept of wrapping the carriers around the cans as they rode down the conveyor lines, raising the production to over 900 per minute. Then, like an endless seesaw, C.C.A. had to go back and meet that competition with more ingenuity to hold its place in the field.

The Klik-Lok Tray Success Story

The Klik-Lok Corporation developed several packaging systems based upon a simple but very useful paperboard tray. By inventing machines for taking the tray blanks, one at a time, and locking their four corners simultaneously, these Klik-Lok trays became widely used by makers of baked goods, precut meats, greeting cards and many other diverse kinds of merchandise. The machines made the trays the useful package they are today. However, the tray did have to be created first, and with mechanization in mind, right from the start.

A common error made by inventors (including the writer. in his younger days) is to be retained to design a package for some corporation, do it and find a widely useful type of construction resulting from the project, but parting with all rights to it, for a price, not realizing that many times that much could be realized from its sale to others. Usually a big company like G.E. would want to use a development for only a very limited part of its potential. Later, some more perceptive man turns the one-shot design into a broad business. The Klik-Lok Corporation had a broader, more sophisticated viewpoint.

There are more success stories in packaging than we could tell here . . . some great, some small—for this book, however, it is sufficient to say that those who made the most from their ideas were the ones who got out and sold them. Many companies were started around such simple ideas of packing as how to blow open a static-stuck plastic bag so it could have garments slipped into it. While very simple, it was based upon *seeing* a need firsthand and thinking up a good solution for it. Being *in* a business is the stimulating way to hatch out ideas for that business . . . not in an "ivory tower" at some remote point, like your home.

For those who would like to learn more about this excellent field, here are some starting points .

Where to Get Information on Packaging and Packing

Because the field is so enormous, no one publication is enough to cover it; however, one very fine source of information is the *Modern Packaging* magazine and the *Modern Packaging Encyclopedia*. The publisher is Morgan Grampian Inc., 2 Park Avenue, New York, N.Y. 10016. The magazine subscription is $20 per year (*Encyclopedia* free with it). Also, such magazines as *Paper Board Packaging*, *Package Engineering*, *Modern Plastics* magazine and many others can be seen at the New York Public Library, Main Branch, Magazine Room. The *Modern Packaging Encyclopedia* is your guide to just about all the suppliers.

The *Package Engineering* magazine is published at 2 North Riverside Plaza, Chicago, Illinois 60606, and the subscription price is $10 per year.

The *Paper Board Packaging* magazine is published by

Magazines for Industry at 777 3rd Avenue, New York 10017, and the subscription price is $ 11 per year.

For plastic packaging, both the monthly magazine called *Modern Plastics* and its *Modern Plastics Encyclopedia* correspond in plastics to the *Modern Packaging* source listed in the preceding paragraphs, and they have the same publisher, address and subscription price. These are excellent references for that trade.

* * * * *

THE PACKAGING MACHINERY FIELD

The field of packaging machinery presents a vast, endless area of diverse opportunity to both the inventor and the inventor-businessman. In a nation where mass production is the very keystone of our economy, packaging machinery becomes the means by which mass-produced goods can be made at prices low enough to attract the volume necessary to run our factories.

Opportunities in Packaging Machinery

From the viewpoint of the creative-minded, one element becomes the dividing line between the areas open to outside inventors and those areas which are confined to "house talent." That line delineates for the outsider those machines and devices which are of *general useful-*

ness . . . which can be sold broadly. The insiders are usually concerned with developing a machine for erecting or making some single patented package, and the job is commissioned by either the package manufacturer or the package user, (packer). Custom-made machinery of this type is not marketed by the inventor; it is part of a "packaging system," and, *if* it is marketed at all, it is by the package supplier.

Because of its enormity, for us to say that packaging machinery is a promising field of endeavor may be akin to saying that the United States is a land of opportunity. True, it is. But, for one, that may mean a chance to own a Florida orange grove . . . to another, an opportunity to give a piano recital at Lincoln Center. The extremes of packaging machinery are indeed as diverse and as attractively remunerative.

Opportunities for the "Insider" vs. the "Outsider"

Let us select, at random, typical inventive projects which would be handled by these two opposing creative talents. With a yardstick of "general usefulness," we measure a good item for inventing and selling broadly to almost anyone—a device for ejecting (dispensing) typical gummed paper tape for sealing a package, the new device making it possible to offer the small shopowner a chance to print his name and advertisement upon the tape as it is ejected. This could be sold to almost any retail shopkeeper; therefore, it would be a good item for the *outsider* to develop.

In contrast, a machine for automatically setting up a six-bottle carrier for loading at a bottler's plant, is tied to one specific carrier; it is part of a packaging system—package and machine combination—that is sold or leased as a "deal" by the package supplier (box house). Such a machine is developed to order for that supplier and may be the product of his own machine shops (the largest firms only), or it is given to some outside, independent development group who not only can develop the original machine, but also can duplicate it, as required, for the many customers the supplier sells on his "system." Even if the outsider has the opportunity to invent such a machine–*unasked*–he should avoid it, lest he become "boxed in" with one customer who may not even want his invention.

The outside machinery supplier who is under contract to do a job for the packaging supplier, is not to be compared to an outsider such as we refer to, who invents a machine–*unasked*–and then has to go through a most difficult selling job to get his money out of it, with or without a profit. There are certain jobs undertaken on a more flexible arrangement, and that is well worth studying as a more acceptable alternative.

When a certain well-known brewery was dissatisfied with the complexity of their packing machinery for six-can carriers, they had a company develop a simpler, better piece of equipment. The designing firm was given the privilege of selling the machines to other brewers, too, and for that consideration, made a much lower price for the original development than if they could have sold the machines only to the original brewer.

With such machines as we are discussing, running into the $100,000 class, the special "consideration" could be a substantial savings on the original prototype. The brewers or bottlers who take on the new system buy their own machinery; hence, they are not of as vital a concern to the man who sells the system. The machine designer can make up his profits on the packer (user).

The outside development houses, such as the one referred to above, prefer to go into projects with high sales potential for their machines; even if the supplier pays very well, they do not want to get into a heavy designing job, only to sell one machine. These men are very well paid as salaried employees, or even small machine manufacturing company owners, but, they do not amass great fortunes from their creations. Conversely, they do not devote endless hours in unpaid speculative inventing; take great financial risks; spend money for patents; devote endless days to selling. Once the new can-packer was developed, the original package supplier took on the selling of the machines–machinists seldom sell!

Only those who take on financial risks make the "big money" in inventions, and it is the *manufacturing* which produces the dollars, *not* the inventing. Why is this? Because big risk can mean possible failure as well as possible success. The machinists were *hired* to do a nonrisk job.

The Need for a Working Model

It is the custom of the machine buyers to want to *see* the machine they intend buying before purchasing it; they do *not* like the idea of buying blueprints or plans, for that places the responsibility of the correct performance upon them. Even the custom-made machines must perform at a predetermined level of output and quality to be acceptable. The inventor who approaches a prospective buyer with plans or ideas alone better be a fantastic salesman. The best he can hope for is partial financing.

For the outside inventor, the most promising area for his efforts are the *general needs* which have developed so broadly these past few years: the many, many forms of shrink packaging and shrink packing; the packing aids for plastic bag packing as used in the shirt, sweater and lingerie factories; the supermarket packing of fruits and vegetables in predetermined quantities; the packing of large appliances and kitchenware in clear, plastic shrink-film packs. Such developments cater to the needs of *many* packers, and can be sold without requiring the license of some patentee.

One inventor saw the disadvantage of trying to open clear plastic bags stuck together by static electricity. He made a simple, ingenious, air-opened bag device, which not only opens the bag but holds it that way while the merchandise is being inserted. Many softgoods packers needed that!

While most amateur inventors tinker with the machinery they are trying to invent, creating it piece-by-piece by trial and error, the big machinery houses make precise plans and build the machines from those plans. The amateur inventor is amazed to learn that a machinery manufacturer cannot make a model from a model; he can only work from plans. In one case, recently, it cost over $10,000 to make the *plans* for a big printing-and-cutting press (which came from Germany); the press's owner thought he could have his *press* copied. He could not.

The Big Corporation Success Stories

The many inspiring stories of past successes in the packaging field can be very briefly reviewed here, but each is worth far more time than this volume permits for such a purpose. The Peters Machinery Company of Chicago has built a very prosperous business in making machines to set up biscuit cartons (ala Uneeda Biscuits); Hayssen Bread Wrapping Machines are now used for all sorts of plastic film wrapping and wax paper wrapping, with the electric eye registering the printing correctly on the package.

Egg box erecting machines and the egg boxes themselves, are produced by several large box houses: Mead Corporation, Container Corporation of America and Diamond National, among others. The same field also supports the molded-pulp trays from Keyes Fibre and many new forms of heat-molded, off-a-roll plastic boxes which are *very* light and strong.

Each of the successful developments just described had resulted from the work of *inventors in the field.* Even for them, it was not easy . . . but it *was done,* and successfully.

Even the general ideas about the handling of the shrink-film came from men who *saw* the need; they did not think it up themselves. Packaging has enough branches, varieties and degrees of complexity to provide opportunities for everyone, but only those who carefully observe their potential areas of development and then patiently search the patent office to insure their safety, can hope to avail themselves of those opportunities.

Where to Find Further Information

The trade journals of the packaging field are themselves diverse. The leading *packagers* of merchandise read *Modern Packaging* and its sister publication, *Modern Plastics;* they also depend upon the remarkable, thick-as-a-phonebook *Packaging Encyclopedia* (given free with a subscription to *Modern Packaging* magazine; the same deal is made for the *Modern Plastics Encyclopedia* with the *Modern Plastics* magazine subscription). Shipping box needs are found in *Material Handling* magazine and also in *Package Engineering,* monthly. Paper cartons and boxes are treated for the packaging maker and engineer in *Paperboard Packaging,* monthly. All can be found in the New York Telephone Book and are available almost everywhere.

A most valuable Packaging Machinery Catalogue is available, published by *Package Engineering* magazine (see prior information). Most of these catalogues, encyclopedias and magazines are available in the big-city public libraries. There may be seen the impossible task of reviewing the field without doing injustice to many of the worthy innovators in the forefront of development: the Klik-Lok Corporation, with their cleverly designed trays for anything from greeting cards to frozen foods, and the machines for erecting them automatically; The Stokes and Smith "form and fill" machines, which form a continuous length of plastic tube into candy-filled plastic bags, with the candy dropped in as the bags are formed; etc. Each development, attained as a step-by-step job, from package concept to machine concept, was as much a victory in the business acumen of getting the whole complex project "on the road" as getting the inventive components functioning.

While the author has made plain that he is enthused by the breadth of the field, the *ever changing* needs which foster new ideas and inventions and the inevitable dollar considerations, everything does come back to the one premise which means either success or failure in *any* field . . . creating from a background of *knowledge,* firsthand knowledge of one's chosen field.

For those who can afford to attend the packaging machinery shows, the actual value of the demonstrations to be seen is incalculable. The machines are set up and running, and they are open to inspection, questions and explanations of the most technical kind. Obviously *no* photographs of a machine in a magazine or catalogue offer what a firsthand view of the machine in action can offer. Getting a clear idea of what now exists in a given field can indeed save many a costly misstep for any inventor.

At trade shows, much valuable literature is available for home study, too. Remember: The packaging machinery field covers so broad an area that no one man can hope to "specialize" in it. Get oriented toward some finite division of packaging, if you would learn enough about it to attempt a new packaging machinery concept.

CHAPTER FOURTEEN

THE IDEA MARKETS: Additional Fields of Opportunity

People do not have "needs" for home appliances. They are used to their accustomed ways and even their inconveniences, and it takes years of hard selling to switch them. Inventors and their sales representatives turn people's inconveniences, troubles and nuisances into "needs," and then must convince people that they really *need* their devices and will come to like them better.

The Electric Appliance Field

PERIODICALLY, the electric appliance field rises to a comfortable plateau, and hovers there, unchanged, long enough to give the illusion that all legitimate needs have been satisfied. Then, overnight, along comes some inventor to open new "frontiers," with such items as the Water Pik electric toothbrush, the Hotray, electric pencil sharpener, electric fry pan, instant-perk coffee-maker, and . . . the field is off and running again, onward and upward in sales, size and opportunity.

How simple it is to see the direction for one's efforts *after* some ingenious inventor shows us the way. One of our most able idea men, Raymond Eisenhardt, had a line on the bottom of his letterheads:

ALL PROGRESS STARTS WITH AN IDEA

I never forgot it, and neither should you. A field which seems to offer no new opportunity to some up-and-coming inventor can sometimes be transformed into "a land of golden opportunity" by a thought!

The Need for "Ground Breakers"

Far too many outside inventors think in terms of improving existing appliances instead of originating new ones, and while there is an undeniable need for ideas for bettering existing appliances, manufacturers can usually get that type of idea from their own design sources. Effort spent improving existing devices is not rewarding to outsiders.

Casual observation will tell you that the well-established manufacturers of home appliances concentrate on *volume*... and that means holding to the solidly proven, broadly acceptable devices. Despite this obvious truth, inventors persist in bringing to them new-need, still-to-be-proven appliances which require costly sales missionary work and bring uncertain profits. Then these inventors are disappointed at their lack of receptive enthusiasm.

When one studies the sales procedures of the successful marketers of *new* appliances, we find that companies were created to market them. *Analytical* sales efforts were necessary to put them over. Typical of the newer success stories is the dental appliance, the Water Pik, a formidable contender for the toothbrush market, especially suited to wearers of dentures.

Instead of trying to sell the public "cold," sales management sought the approval of the dental profession;

with personal visits, patient demonstrations and tact, they earned the recommendations of the dentists, and with that backing, supplied the retail outlets in high-class drugstores. As a "high ticket item," the Water Pik was a welcome newcomer, and has gotten to be a profitable, well-received appliance. The inventor of the Water Pik indeed broke new ground and found rewards for doing so.

The Importance of Personal Demonstrations

Only patient, personal demonstrations and sales efforts put over the electric broilers, electric fry pans, instant-perk coffee-makers and Hotrays. Later, when acceptability was a conceded fact, the stores sold from displays only—but that took a long time. It was the *years of small-company-type sales efforts* which *made the success of a device possible.* A lifetime ago, the now-used-by-everyone vacuum cleaner—big and heavy as it was then—was dragged around from door-to-door, to make sales. Such marketing may be a real cross to bear, but it pays off.

Pioneers may sweat out the early years, but they also establish themselves solidly, and one day become the "big names" of their industry. The field of electrical appliances is an ideal area for establishing a new business, because the buying market is without limit; items are sufficiently high in cost to warrant the interest of the retailer, wholesaler and manufacturer, and the use of labor-saving devices is steadily increasing.

The unique "ground breakers" actually lend themselves to the founding of a new firm even more than an improved standard appliance, for, if you develop an improved device, all you can hope for is just to take a minor position in a big-name field, and that means trading on low prices or secondary sales features (the acceptability of the appliance already being established).

If you are fortunate enough to have invented a truly

unique device, yet you are unable to found a new marketing firm, greater opportunity may lie with the smaller, but still affluent appliance manufacturer who is "on the make." Another choice is an individual with sufficient capital and ability to found a firm with you as a partner. Many an inventor who could not find the proper backing to start a business, went to an investment banker, such as Kuhn Loeb & Co. or Lehman Brothers, and they were able to bring him together with the people who were looking for just such opportunities.

The Possibility of Foreign Financing

The author, when unable to market certain of his own devices in the United States, found the needed backing in Canada, England, Holland and Germany. Japan is good for electronic devices, but negotiations are complicated and should not be attempted alone. Japan has people in New York City to help you, and they speak excellent English.

For anyone contemplating the electric appliance field as a possible outlet for his inventive ability, or as a market for an already invented device, his attendance at a housewares show is an absolute "must." These shows are announced long in advance, and the date is publicized in various trade magazines such as *Merchandising Week* (formerly *Electrical Merchandising*), available by subscription–$10 annually. Write: Merchandising Week, 165 West 46th Street, New York 10036. They are also announced in *Modern Packaging* magazine (see page 150).

As advocated previously, reading the trade papers of the industry is also a "must." Not only does it give you proper perspective of what the best brains in the industry are doing, but also it can spark your own thinking and give it direction.

The previous chapters of this book have covered the

areas of marketing and financing which apply to electrical home appliances, and, when the field is of more than casual interest because of a finite idea or invention, those passages should be read anew, with more perception, in view of your specific application for them. Each *area* of this field requires separate study; the trade details affecting electric irons does not apply to clothes driers, despite the fact that they are both home laundry items.

* * * * *

THE JIGS AND FIXTURES FIELD

America has always been rich in ingenuity and talent. Much of that talent stems from people who have a natural ability to solve complex mechanical problems, but who lack the prestige of college degrees and formal education. Fortunately, there are fields where such men are welcomed and appreciated. This is one of them.

The little-known and not properly respected field of production Jigs and Fixtures is primarily an area of employment, and its opportunities lie with manufacturers of merchandise which require hand assembly of component parts. The designer of jigs and fixtures invents, to order, ingenious ways of speeding up the work, making it more convenient to handle and thus, reducing the cost of producing it. These devices can be extremely simple or very complex, but *all* have one quality: they substitute ingenuity for hard work, and in doing so, speed up production.

In a lifetime of factory production experience, the author has found ingenious jigs and fixtures men a rare form of blessing. No college course can guide them through the trackless wilderness of inventing new devices for producing items never before attempted—they must instinctively know the way. Lucky indeed is the company who has a good jigs and fixtures man, for it can toss a difficult assembly job onto his untemperamental "lap" and know that the answer he will come up with will be practical and efficient.

One of the typical production men with whom the author worked in the assembly of one of his devices, disregarded the advice of the company's engineers, who suggested a complex way to do the job, and reduced a three-girl assembly requiring skill, to a one-girl assembly requiring none. Today, the good "blue collar worker" is the highly paid "prince of production," he wouldn't change places with the "white collar worker" for anything.

True, none of these able workers became famous; they just made a very good living selling ideas and inventions, for good weekly wages . . . no sales problems, no patent costs, no negotiating for every job's selling price. Because such ingenuity is indeed *un*common, it is the basis for hiring—the important thing here is *not* whether the applicant has college degrees, but whether he can solve production problems.

Job Titles in This Field May Mislead You

Frequently jigs and fixtures men have misleading titles, such as foreman of handwork, production supervisor, etc., because the men who decide how assembly and packing is to be done are the ideal people to teach it to the ones who are to do it.

Types of Industries Requiring Jigs and Fixtures

To analyze which forms of manufacture require men to devise jigs and fixtures to aid their production, it is best to seek *first* those who make "one shot" production runs, where each job is different and will have to be set up anew. We may start with the field of advertising displays; here display devices which may require simple folding and assembly, and then packing in cartons, will be followed with a display which is motorized, lighted and assembled out of half-a-dozen pieces. The packing alone is a big undertaking. In displays, *every* job is different . . . every job needs a man to set it up for production.

In greeting cards, too, the hand-pasting and folding of novel cards is a constantly repeated procedure . . . always new, always different. Handbags for women, too, have constantly changing designs, and each bag has to be set up for efficient production. Toys usually sell for *one* year; then, they may possibly be repeated the next year if they are still selling, but far more frequently the next year brings new items, and, of course, new production problems. Toys are indeed a giant field for jigs and fixtures men.

There are a few machine shops which offer a free-lance jigs and fixtures designing service for those who need such devices rarely, or at least not regularly enough, to employ a man for them. Most jigs and fixtures men do have regular jobs, some serving two noncompeting firms.

When a jigs and fixtures man does bring an improvement to a machine of general use, as he may do after seeing some of the problems involved, then he must return to the marketing methods of Chapter Three, to arrange proper remuneration, and, if appropriate, the rights to market the device or improvement outside the firm in which it was developed. It is always preferable to make arrangements *before* doing the work; if anything is wrong, learn about it before you risk time and money.

Working as assistant foreman, or even factory worker, is the background which allows observation, firsthand, of the problems of that particular industry, and leads to a regular job making production jigs and fixtures. Incidentally, the devices which are designed, depending upon the field, can range from a very simple piece of wood, strategically placed, to a metal fixture made in a machine shop for aiding a handworker to do some too-complex a manual job. Designs are literally without end.

Figuring an Estimate

One of the most important elements of a jigs and fixtures man's work is to help a firm figure an estimate to *get a job,* for they frequently cannot estimate the handwork cost until they know how it will be handled. Jigs and fixtures men are indeed an asset to any team in manufacturing.

* * * * *

THE FIELD OF CORPORATE RESEARCH AND DEVELOPMENT

Corporate research and development strives to make scientific progress a product of teamwork in a stimulating, protected atmosphere. No one can deny the value of its premise. However, no stimulus, no protection, can ever replace the inspiring effect of a pile of unpaid bills to a young, married inventor.

The large corporations have spent long years and much of their stockholders' money on building up Corporate Research and Development Departments, and corresponding sales-launching facilities known as the New Products Departments. They are both a most necessary adjunct to big business, and to deny their value would be ridiculous.

Corporations may use product ideas from the outside, *if* those ideas are of sufficient stature or such worth that the corporation would benefit greatly by their use. The truly unique makes its own rules, but no run-of-the-mill inventors or their ideas would be welcomed here.

The educational standards set by the corporations for their research and development personnel are high indeed, and anyone with a B.S. in science would find it of little help, unless accompanied by very high marks or exceptional talent. Many a corporation has sent a young man back to school for higher degrees to qualify him for their type of high-specialty work. Yes! This is indeed the place for rich rewards—as *salaries*—for the qualified.

Your Chance for Success as a Salaried Idea Man

The great American Dream to become rich has many widely different methods of attainment. If you decide to put your inventive talent to work for a corporation, certainly getting a handsome, steady income (and having the good sense to invest it properly) is even better, in many cases, than striving long years on little, awaiting the "big killing," which may or may not come. The money paid to salaried researchers, together with the retirement plans and, later, stock options, has much to be said for it.

This book has been directed toward the independent, talented inventor, as contrasted with the scientifically trained man of many degrees, because the latter essentially doesn't need the kind of help this book offers. However, let us realize that *many forms of research cannot be done by the independent inventor.* When Bethlehem Steel Corporation must develop a new steel to overcome the deleterious effect of rust, no clever inventor is going to cook up a batch of steel in his basement workshop. Some things *are* big and must be handled by big, well-equipped facilities. Outsiders just do not belong.

In good times and bad, there *are* opportunities for qualified, technically trained people. Despite war-to-peace changeover, in certain areas of our country there is a never ending quest for specialized engineers, scientists, inventors and other hard-to-get help. See the proof of this *every* weekend in that great talent market, the Business Section of the Sunday *New York Times,* where as many as 20 or more full pages of display ads reach out for talent in the $15,000 to $50,000 range; these companies even send their executives up to interview sessions at the midtown New York hotels to seek out and hire the most promising.

In corporate research and development, the policy is to reward with salary commensurate with ability, experience and education. Lump-sum remuneration is practically non-existent, as previously stressed, because lump-sum payments are for outsiders selling inventions. The results of research and development work are considered amply paid for by salary. Frequently, salaries, over a period of time, exceed lump-sum payments.

In New Products Departments, there is a greater need for idea men than inventors. Here they pass on potential products from outsiders as well as dream up custom-made products of their own. When the idea man's ideas are good, research and development is given the job of taking a nebulous thought and giving it substance and practical application. Obviously, proper education is necessary; talent is expected, too. The man who puts his energies to work along lines which are suitable to his ability saves himself a lot of useless motions and gives himself time to pursue attainable goals.

Many Opportunities for Creativity

Research and development spans many fields, and today there are excellent departments established in such diverse fields as: baked goods, steel metallurgy, aluminum metallurgy, computers, telephony, electronic printing, cosmetics, plastic films, injection-molded plastics, blow molding, package development, frozen foods, freeze-dried foods, pharmaceuticals, underwater exploration, containerized ships, sea lane oil transportation, refrigeration. This list is actually no more than a tiny fraction of the whole, but it does show that research and development can be as diverse as crackers and submarines, with opportunities for specialization by the qualified.

Despite the fact that unpaid bills have inspired many an inventor to bring forth inspired thinking, it would be unfair not to recognize the premise that a good salary never stopped anyone from being a genius.

CHAPTER FIFTEEN

HOW TO PROTECT YOUR INVENTION

The inventor often has a sense of being protected when actually his so-called protection is worthless. In this chapter, we shall specify the kind of protection available to you, and how to make sure you really *are* protected. This is indeed the basic knowledge of your trade, so learn it well: precisely where to go for information; whom to write to; whom you can reach by telephone.

THERE has always been much misinformation guiding the creative people in our society; most of which has resulted from the tendency to assume, rather than dig for facts. Such a tendency has proven detrimental to their best interests; hence, it is the purpose of this book to provide a readily accessible source of factual information, deliberately simplified, but sufficiently complete, to show the right way—guide the inventors toward where the additional necessary details may be found.

Get into the habit of *double-checking "facts"* lest they be obsolete "facts" which are no longer correct. Patent offices move; fees go up; laws change; procedures change. It will pay you to *check!*

What Forms of Protection Are Really Necessary

A basic essential for the protection of any invention is to have it *recorded* in writing, in detail, witnessed by two people . . . clearly dated and signed by the inventor. If the invention can utilize drawings for clarification, they too should be dated and signed. This whole procedure is relatively simple in concept, yet, due to misinformation from unauthorized or ill-informed sources, and knowledge of now-obsolete "facts," it is necessary to restate the correct procedure here, in the form recommended by the U.S. Patent Office.

Information on Registering Dates of Conception

Creative people tend to be secretive about their creations and their work. While secrecy is necessary prior to the issuance of a patent, unfortunately there is no protection in it, for others may well arrive at the same solution to a problem, having been exposed to the same stimulus. That is why you should seek patent protection as the *only* true protection available to you. Registering your date of conception is the first step toward that goal.

While it is *still legal* to make a written record of invention, have it witnessed, signed by the inventor, and then signed by a notary public, a much better way is now available to you: a way that cannot be doubted for its authenticity . . . an *official* record.

Because witnesses and even "friendly" notaries could be doubted, the United States Patent Office has at last set up a much-needed service: You can send your Record of

Invention to the Patent Office, *in duplicate,* and they will date both copies officially. Then they will keep *one* copy and return the other to you. Why *two* copies? Because *they will keep this record for only two years.* If you apply for a patent, it will be added to the application, but, if not, their copy will be destroyed.

It *is* often necessary to delay (mostly to improve) your application for a patent, so having your own copy of the date of your ideas, which cannot be questioned, is a far better proof than the older method. To learn the details, send for the *free* booklet entitled: Disclosure Document Program to: The U.S. Department of Commerce, Patent Office, Washington D. C. 20231.

This is *not* a "caveat" or temporary patent such as the Patent Office issued in former years. It is just an unquestionable record of the existence of your idea when the date states it was stamped by the Patent Office. It carries *no patent privileges!* You cannot stop others with a Record of Invention. You *may* insure yourself a patent, someday, with such a record. A patent is a monopoly; until you *are* granted a monopoly, you have no right to exclude others from making or selling your idea.

To prepare an acceptable Record of Invention properly, use a good-quality white paper of a size not to exceed 8½" x 13" (regular 8½" x 11" letter size is acceptable), and type a brief, lucid explanation of your *invention* (*not* personal details as to where and how you got the inspiration).

Describe the *structure;* its function; its intended uses and its essential differences from existing related devices. If the invention can be illustrated in drawings, use letters or numbers on those drawings to denote important points or areas, and use those references in your written explanation. If you must write by hand, use *ink, not* pencil, for pencil smudges with the years, and can be erased and changed long after the sketches were made; hence, it is not legal. Sign and date your disclosure; have two witnesses read and sign it too. For a duplicate, have a clear Xerox copy made of your Record of Invention, and for safety, have two made up (one to keep until the dated record is returned). Then, send the original and one copy to: Commissioner of Patents, U.S. Patent Office, Washington D.C. 20231, together with a $10 check or money order (do not send cash) made out to the Commissioner of Patents. If more than one page of disclosure is necessary, number each page thus: page 1 of two pages (or three), and sign and date each page at the bottom. Now, send the two copies with a self-addressed envelope, properly stamped, for their return of your duplicate copy. Attach a separate note reading:

> The undersigned, being the inventor of the disclosed invention, requests that the enclosed papers be accepted under the Disclosed Document Program, and that they be preserved for a period of two years. Please stamp-date and return the duplicate copy.

Simple though this procedure is, many inventors have failed to follow it, and their "wonderful idea" remained just that . . . an idea–a nebulous thought in their heads, *not* on paper. The Patent Office does not recognize thoughts as inventions, for no one can testify dependably to their existence nor dates. Even the existence of experimental models cannot, of themselves, prove dates. Dates are proved by *writing.* Develop skill in making believable records of your ideas.

Several times, over the years, the writer has won interference proceedings between two or more inventors, each claiming to have invented the device first. He won because his records were clear, properly dated and properly witnessed and notarized. The others had pencil sketches (sometimes on the back of envelopes or bills) and had kept these secret . . . not seen by anyone but the inventor. Considering that a notary public is available in every town and the notarizing costs so little, failing is inexcusable.

Be sure your witnesses read and understand the disclosure sufficiently to testify, in court, *if* necessary.

What Registering Your Date of Conception Does for You

While it may seem obvious to professional inventors that no privileges accrue to those who have offical Records of Invention, except for proof of dates of conception, perhaps a more detailed explanation should be forthcoming for those who are neophytes. The reason that witnessed Records of Invention are not monopolies is simply because such ideas and inventions as may be involved may well be only duplicates of existing patents, or already expired patents, or may not even be adjudged sufficiently ingenious to qualify as "inventions." *Only* the Patent Office can award privileges, and those only after long and careful searches and consideration.

Now, with an officially dated Record of Invention, can the inventor sit back and feel protected? *Definitely not!* Why? What if two equally diligent, equally inspired inventors create the same thing? To whom should the patent be granted? Both saw the same need; both arrived at the same conclusion as to how to satisfy it; however, one was a week or a month earlier than the other. Does the inventor who was "the first to conceive the idea" get the patent? *Not necessarily!* There are still a series of other qualifying acts which must be brought into the picture. It is most important that you learn them well.

Here, if one has an official Record of Invention and the other only ideas and sketches he made and dated himself, the witnessed documents must precede the self-witnessed records in the eyes of the Patent Office examiners. What if both have the same date and both have witnessed or Official Records of Invention . . . what then?

The next qualifying requirement brought into the case would be centered upon the diligence displayed by the inventor. Did he make an actual model? In view of the fact that a great portion of the ideas recorded as words and sketches would be inoperable if transformed into working models, he who did make a model which did operate as claimed, would be awarded a patent over he who merely wrote about it. The Patent Office refers to this as "reduction to practice."

If, as has been the case in the writer's own experience, three people did have good records, did have them duly notarized, did make models and did make them work, what then? Well, the writer was two years ahead of the others in conception, a little behind the others in date of filing the application, but was undeniably first. Did he get the award? No! Why? Because he delayed unduly in filing and was considered to have "abandoned" his invention and later to have revived it. He claimed that during World War II the government itself forbade the use of the invention (bottle-carriers), so he was justified in not filing then.

The examiners ruled that the others *did* show the proper diligence in filing for a patent; the writer did not. His two-year date exceeded legally allowed "time before filing." He was really just "keeping it secret," which never carries any protection for the inventor. Without close interference he would have gotten the patent, for the subject would not have even come up.

The reason for the patent examiners favoring those who make models over those who only write about them is best demonstrated in this example: If several inventors conceived the idea of eliminating office building elevators by some proposed device which would enable one to step from his office window and float gently to earth, *only* the man who could make a model that worked would deserve consideration—the others merely had an idea . . . a nebulous dream. However, while it is in the realm of possibility that someone could invent such a device, and a broad patent would issue for it, would that mean that no one else could patent such a device, too?

Here we must make the distinction between patenting the whole broad concept of the machine for floating to earth and patenting the particular physical embodiment given to it by the inventor. Concepts cannot be patented! If ten different, *noninfringing ways* could be designed to do the same job, then ten patents could be granted. One could be a jet-stream device, another a parachute, another a pole-slide with controls. Thus no two *physical* embodiments would interfere, but the concept would be used by all. The Patent Office must keep the field open for the ingenious; hence, no extremely broad claims are issued.

One final factor, mentioned previously, can be an important point to keep in mind. Many people who create try not to spend their money for patent applications until some buyer seems interested. Such an inventor may have a search of patents made and then relax and feel safe. However, no one may search the *pending patents;* hence, a patent may be issued after that first patent search has been completed, then, when our inventor applies for his patent, he is shocked to learn that his device is no longer patentable. Moral: If your device is worth patenting, don't delay.

Protection by a "Diary of Progress"

There is a more detailed version of making proper Records of Invention. Companies which develop inventions commercially keep a regular running record of those activities and advances. Unlike the one-shot inventor, they keep a regular diary of progress, knowing that such a book may one day appear in court. *When* possible, even these continuing developments should be periodically transformed into the new government-approved and dated documents. The book method, still in use, is based upon a *bound book, not* a loose-leaf book, and depends for its truth on the dated entries before and after.

Many young inventors seem to feel that the Patent Office strives to make life difficult and unnecessarily complex for them. Experienced inventors get to know that the patent examiners actually strive to *aid* them. When an examiner rejects claims, it is not to hurt the inventor, but rather to make him answer why his invention should be granted a patent despite the existence of what appears to be a conflicting device. This places the answers in written form in the public records, and later, if the case is allowed, no one can come forth to claim that had the examiner known of the existence of some pertinent patent, the patent to the inventor never would have been allowed. By *giving the inventor this opportunity* in his handling of the case, the examiner is insuring a much stronger patent than if he tried to favor the inventor by not requiring answers to those references. There is no spite work in the Patent Office.

The Cost of a United States Patent Application

Inventors frequently ask about the cost of preparing a patent application and are distressed by not obtaining a simple, straightforward answer. Actually, patent attorneys set their fees both by the time involved as well as the size and importance of the client—there is no one, set rate. The major cost of the application lies in the attorney's work, and that includes much time spent talking to clients, thinking about how to proceed and other necessary but intangible services. Attorneys who deal with creative individuals find that they must set a finite figure for what may be an infinite amount of work. They try to "cover themselves."

Most regular business accounts allow the attorney to do the work, then, *knowing* what time he spent, send an appropriate bill. An unethical attorney may "pad the bill," but such tactics are primarily used on very careless neophytes. The writer found that attorneys were, with rare exceptions, both honest and ethical. The better attorneys would not quote a fixed charge, because such a charge had to cover too many unknown factors. Fixed-priced patent attorneys cater to an inexperienced, amateur clientele.

In contrast to this, the government publishes very clean-cut lists of fees and costs for all items relating to patents and patent applications. In their free booklet entitled General Information Concerning Patents, obtainable from the U.S. Department of Commerce, Washington D.C., you will read answers you can understand and depend upon. For your convenience, these government figures are given on pages 182-187. Depending upon when you read this list, it may be well to check the prices by sending for the latest copy of the above-mentioned book, for prices are subject to change.

The writer acknowledges the dangers inherent in daring to quote attorney's fees, yet, he realizes that many inexperienced inventors do desire some "ball-park figure." It should be stated that all of the writer's experience has been with big-city attorneys, and he can only surmise that a small-town attorney may be slightly lower in his fees, due to lower rent and overhead.

The following figures are approximations of what the writer believes a reputable attorney might charge, based on his own recent, innumerable applications.

For a preliminary search on so simple a subject as a manual can opener, a new paper clip, a phonograph record rack, a novel cuff link or ash tray, the fee would run from $100 to $200 depending upon how thorough a search seems to be indicated. *If* the device is slated for immediate production, a thorough search is advisable. Search fees are split with professional searchers in Washington D.C. Do *not* ask your attorney to do the searching personally—the professionals are best.

On the above-mentioned simple subjects, a reasonable fee for the attorney, plus the draftsman's fee, plus the government filing fee, may run between $500 and $800, total.

The same attorney would handle a more complex case such as a postal scale, an electric broiler, an automobile jack or a simple mechanical toy for between $500 and $1,200, total.

Going past this point to computers, knitting machines, jet engines and other complex mechanisms, the cost would run into many thousands of dollars for the preparation and far more for government fees, especially as the applications run into many, many pages of drawings, specifications and claims. The printed price list will make this subject far simpler to estimate. Attorneys' fees still must take into consideration the man's ability and your own financial position.

After a period of from eight months to as long as a year and a half, the Patent Office will act upon your application. This so-called Patent Office Action must be answered with an "Amendment" composed by your attorney (preferably with your help). This Amendment will vary in length and complexity according to the nature of the case. For our ball-park figure, the simplest category may require an Amendment costing from $100 to $300. The next category up may run from $200 to $500. The "heavy" category could vary far too widely to estimate.

Despite the fact that most inventors have very limited budgets, they must try to avoid forcing their attorneys to quote flat prices, because attorneys cannot possibly estimate, in advance, what difficulties they may encounter in preparing the case; studying out the views for the draftsman; discussing the case with the inventor; studying the references brought out by the search and doing the thousand and one things which come up *after* they get into the case, but which cannot be seen at the beginning. It is best to choose a good, honest, reputable attorney and trust him, rather than go to a "flat-price man" who may give you a quick, inadequate patent which leaves you unprotected.

Perhaps the most important element of the application lies in having before you all the prior art, uncovered by a good search. A really thorough search costs more than most creative people wish to spend. However, the writer cannot stress too strongly the need for every inventor to do his own searching.

For just about what you would have to pay an attorney to make a search for you, you can go to Washington yourself and study the files under the expert guidance of very patient, intelligent men who show you where you may find precisely what you seek. However, with a personal search, not only can you uncover the pertinent art for your attorney, but, equally important, you can see what the best brains in the field have been doing for generations. Low coach fares and innumerable airlines make personal searches feasible. Such a search will indeed broaden your outlook and may well give you a perspective which would better fit you to succeed with your project.

The place to search is no longer in Washington D.C. but rather the new Crystal City, right beside the Airport. If it is necessary to stay more than a day, there is a very nice motel at the Airport, too. However, one day's search is probably twice as long as the experienced professional searcher intends to spend on your project. After a few trips to Crystal City, you too can develop the necessary skills to get all you need in a single day's search. The files remain open until 9:00 p.m. to help you. This "do it yourself" viewpoint may be unorthodox in character, but the writer found it invaluable.

Even an excellent search is not a "clean bill of health," due to the fact that pending patents are not open to search; however, it is as safe a job as you can do prior to the actual filing of the case. When you find pertinent patents, you can get them within a few hours for 50¢ a copy, and most attorneys Xerox them in their own offices to provide the client with an extra copy. Depending upon the number, it is usually best to pay the Patent Office, for they indeed work at nonprofit levels. When they mail you copies, they do not even charge for postage. So, with the Patent Office right near the Airport; with a moderately priced motel at the airport; with all the devices in which you are interested concentrated according to category; with intelligent men there to help you . . . why not try this on your next invention?

Another recommendation is, write for the 30-page booklet entitled For Inventors, by sending 15¢ to: Superintendent of Documents, United States Government Printing Office, Washington D.C. 20402. Besides much helpful information contained in the booklet, it tells you about the many other government publications written expressly to help you.

Despite many legends about the famous inventors of old, no one is supposed to walk into the Patent Office with a model under his arm. In fact, *all* models are expressly forbidden unless specifically requested. Inventions are evaluated from drawings and specifications. The *only* purpose served by a model, as far as the Patent Application is concerned, is to prove that you have a viable, working device . . . not an unproven idea. However, *commercially*, a model will help you sell your invention, as was explained earlier.

Official Patent Information

An attorney resident in the United States and specifically familiar with practice in the United States Patent Office should be employed. An attorney not resident in the United States, if on the Patent Office roster of attorneys, may act, and correspondence will be addressed to him unless he appoints an associate in the United States.

FEES AND PAYMENT

The following fees and charges are payable to the Patent and Trademark Office:

1. Filing fee. On filing each application for an original patent, except in design cases .... $ 65.00
   In addition:
   On filing or on presentation at any other time, for each claim in independent form which is in excess of one 10.00
   For each claim (whether independent or dependent) which is in excess of 10 ...... 2.00
   Errors in payment of the additional fees may be rectified in accordance with regulations of the Commissioner.
2. Issue fee. For issuing each original or reissue patent, except in design cases 100.00
   In addition:
   For each page (or portion thereof) of specification as printed 10.00
   For each sheet of drawing 2.00
3. In design cases:
   a. On filing each design application 20.00
   b. On issuing each design patent:
   For 3 years and 6 months 10.00
   For 7 years 20.00
   For 14 years 30.00
4. Reissues. On filing each application for the reissue of patent 65.00
   In addition:
   On filing or on presentation at any other time, for each claim in independent form which is in excess of the number of independent claims of the original patent .. 10.00
   For each claim (whether independent or dependent) which is in excess of ten and also in excess of the number of claims of the original patent 2.00
   Errors in payment of the additional fees may be rectified in accordance with regulations of the Commissioner.
5. On filing each disclaimer 15.00
6. On filing each petition for the revival of an abandoned application for patent 15.00
7. On filing each petition for the delayed payment of the issue fee 15.00

8. On appeal for the first time from the examiner to the Board of Appeals 50.00
   On filing a brief in support of the appeal .. 50.00
9. For certification of copies of records, etc., in any case, in addition to the cost of copy certified 1.00
10. For certificate of correction of applicant's mistake 15.00
11. For uncertified copies of the specifications and accompanying drawings of patents, except design patents .... .50
12. For uncertified copies of design patents .... .20
13. For recording every assignment, agreement, or other paper relating to the property in a patent or application 20.00
    For each additional item, where the document relates to more than one patent or application 3.00
14. For typewritten copies of records, for each page produced (double-spaced) or fraction thereof 1.50
15. For photocopies or other reproductions of records, drawings or printed material, per page of material copied .... .30
16. For abstracts of title to each patent or application
    For the search, one hour or less, and certificate .. 3.00
    Each additional hour or fraction thereof 1.50
    For each brief from the digest of assignments, of 200 words or less 1.00
    Each additional 100 words or fraction thereof .. .10
17. For translations from foreign languages into English, made only of references cited in applications or of papers filed in the Patent and Trademark Office insofar as facilities may be available:
    Written translations, for every 100 words of the original language, or fraction thereof 5.00
    Oral translations (dictation or assistance) for each one-half hour or fraction thereof that service is rendered 4.00
18. For making patent drawings, when facilities are available, the cost of making the same, minimum charge per sheet 25.00
19. For correcting drawings, the cost of making the correction, minimum charge 3.00
20. For the mounting of unmounted drawings and photoprints received with patent applications, provided they are of approved permanency .. 2.00
21. Lists of U.S. Patents:
    All patents in a subclass, per sheet (containing 100 patent numbers or less) .... .50
    Minimum charge per order 1.00
    Patents in a subclass limited by date or patent number, per sheet (containing 50 numbers or less) .... .50

Minimum charge per order 1.00

22. Search of Patent and Trademark Office records for purposes not otherwise specified in this section, per one-half hour of search or fraction thereof 3.00
23. For special service to expedite furnishing items or services ahead of regular order:
    On orders for copies of U.S. patents and trademark registrations, in addition to the charge for the copies, for each copy ordered ...... .50
    On all other orders or requests for which special service facilities are available, in addition to the regular charge, a special service charge equal to the amount of regular charge; minimum special service charge per order or request 1.00
24. For air mail delivery:
    On "special service" orders to destinations to which U.S. domestic air mail postage rates apply, no additional charge.
    On regular service orders to any destination and "special service" orders to destinations other than those specified in the preceding subparagraph, an additional charge equal to the amount of air mail postage. (Available only when the ordering party has, with the Patent and Trademark Office, a deposit account.)
25. For items and services, that the Commissioner finds may be supplied, for which fees are not specified by statute or by this section, such charges as may be determined by the Commissioner with respect to each such item or service.

The following are sold by the U.S. Department of Commerce, National Technical Information Service, Springfield, Va. 22151, to whom all communications respecting the same should be addressed:

Microfilm Lists of Patents.—The reels containing the mechanical, electrical and chemical patent numbers in original and cross-referenced classification may be purchased for $70.00 or individual reels for $6.00 per reel (identified as PB 185–900). The classification of all design patents is listed on a single reel which may be purchased for $6.00 (PB 185–917).

The following publications are sold, and the prices for them fixed, by the Superintendent of Documents, Government Printing Office, Washington, D.C. 20402, to whom all communications respecting the same should be addressed:

Official Gazette of the United States Patent Office:	
Annual subscription, domestic	$342.20
Annual subscription, foreign	427.75
Single numbers	6.60
Annual Index Relating to Patents, price varies.	
Manual of Classifications of Patents	44.00
Foreign	55.00
Manual of Patent Examining Procedure	19.65
Foreign	24.60
Patents and Inventions. An Information Aid for Inventors	.40
Trademark Rules of Practice of the United States Patent Office With Forms and Statutes	3.50
Foreign	4.50
Patent Laws	.75
Attorneys and Agents Registered to Practice Before the U.S. Patent Office	3.25
Guide for Patent Draftsmen	.65

THE ABOVE PRICES ARE SUBJECT TO CHANGE WITHOUT NOTICE.

All payment of money required for Patent and Trademark Office fees should be made in United States specie, Treasury notes, national bank notes, post office money orders or postal notes payable in Washington, D.C., or by certified checks. If sent in any other form, the Office may delay or cancel the credit until collection is made. Postage stamps are not acceptable. Money orders and checks must be made payable to the Commissioner of Patents and Trademarks. Remittances from foreign countries must be payable and immediately negotiable in the United States for the full amount of the fee required.

Money paid by actual mistake or in excess, such as a payment not required by law, will be refunded, but a mere change of purpose after the payment of money, as when a party desires to withdraw his application for a patent or to withdraw an appeal, will not entitle a party to demand such a return. Amounts of 50 cents or less will not be returned unless specifically demanded, within a reasonable time, nor will the payer be notified of such amount; amounts over 50 cents but less than $1 may be returned in postage stamps; and amounts of one dollar or more will be refunded by check.

FORMS

Forms illustrate the manner of preparing parts of

applications for patent and other papers. Additional forms are given in the Rules of Practice. Forms for patent specifications and drawings are not given since these vary so considerably. Specifications and drawings of patents may be inspected and studied in those libraries which maintain collections of patents. Particular patents of interest can be located from the *Official Gazette* and copies purchased.

How and When to Use Design Patents

Basically, design patents are intended to cover the ornamental aspects of a design or invention. Frequently, the mechanical invention which can be protected by a design patent is of such a nature that the very form of the object controls its use. You *cannot* protect the mechanical aspects of a mechanical device with a design patent, yet, if you were to have a simple invention, such as an arcuate, plastic object, designed to fit a person's shoulder and hold a telephone receiver in usable position against his ear, there is *no* action, nor is there any cooperating mechanism other than the *shape,* the form, the appearance, to do the job. Such an item could be protected by a design patent.

The dress industry, coat, hat and shoe industries all use the design patent to protect their styling. Typewriters, shavers, pens, pen stands, motor boats and many more manufactured items utilize these "protections for the ornamental appearance" despite the fact that ornamental appearance does have certain mechanical functions; thus, while purporting to protect the ornamental appearance, they do accomplish their protection of the mechanical function, too.

In the price list printed a few pages back, it can be seen that design patent government fees are indeed less than the mechanical patent fees; the attorney's fees are proportionate to the government fees because there is so little work involved. Depending upon the subject of your design, protection can be had for three years, seven years or 14 years, and the fee is based upon the time. Unlike the 17-year mechanical patent, the design patent, by its very nature, has a shorter life. The many aspects of fashion merchandise are, of themselves, based upon never ending change.

Many industrial designers cover the smart styling of their appliances and devices so that their *appearance* in the marketplace is protected for their clients. Study the weekly patent gazettes for the wide uses to which design patents can be adapted. They are cheap protection for those who require visual advantages and fashionable angles.

Patent Office Procedures and Specialized Nomenclature

The Patent Office used to allow cases to take as long as five to eight years to be decided. However, today, the policy seems to be to give the inventor two carefully considered Actions, each requiring an Amendment (reply), and then, ready or not, a final decision. The first Action which passes upon an invention and grants the claims is such a rarity as to be almost unheard of. As previously explained, the examiner wants to give the inventor every chance to answer to the pertinent references which he cites. This first Action is given some eight to 18 months after the filing date, and the inventor is given six months to prepare and file his answer (Amendment).

The time you are allowed to send in your Amendment is figured from the date stamped on the face of the Action itself. When you get your second Action, you may be given a shorter time to answer–perhaps three or four months–and this too is unmistakably printed on the Action. If the examiner allows your claims, you will get a Notice of Allowance and will be given a date by which you must send in your final fee (your acceptance of the Allowance). You may wish to have a delay during which you try for broader claims or make other changes, and your attorney has a set procedure for doing this for you.

If instead of a Notice of Allowance you get a Final Rejection, as many of our most successful inventors have in the past, it does *not* mean that you have to abandon your Application. *Many* courses are open to you: The first and best is, to go with your attorney to Crystal City to visit the examiner. An appointment can be made by telephone or letter, but telephone is indeed quicker and better. Then, in a meeting, you can show the examiner a model of your invention (if the subject is appropriate to a portable model) and demonstrate wherein your device serves a purpose worthy of a patent.

The writer has found, many times, that static pen drawings do not portray a mechanical motion or ingenious mechanism; whereas, a device placed before an examiner, and operated for him to watch, makes an entirely different impression upon him. Once convinced of the patentability of a device, the examiner himself will suggest how certain words and phrases may be used to circumvent interference with other existing patents, and he will guide your attorney toward writing an acceptable claim for your invention.

Even past Final Rejection, courses are still open to an inventor, but they run into considerable expense, and should be taken only upon the advice of your attorney. Inventors are sometimes mystified by the obscure words and phrases of the patent examiner, for he never says a simple "yes" or "no." He may say that certain claims "appear allowable"; then why not say they *are* allowed? Because he means that "they appear allowable at this time, but until the case proceeds further, and we consider other later references, a concrete decision cannot be made."

A most frequent misunderstanding is when an examiner is trying to reject *a claim,* while the inventor thinks he is rejecting the patentability of the invention. It is the *wording of the claim* which may be unacceptable, not the invention itself. Sometimes an examiner will try to show that if you took one feature of some older patent, and combined it with another feature of another older patent, you could arrive at your device without sufficient ingenuity being required to deserve a patent for "invention." Here the attorney argues that practically every machine is a combination of old elements that have been put together to produce a new and useful result. However, there must be a degree of ingenuity present which deserves to be called "invention."

Here are many diverse but related subjects which could be decidedly valuable to you in obtaining a patent.

1. Study the claims of the patents of successful devices in your field. Learn the strange "long sentences" which comprise a claim, and the punctuation which makes them a single, acceptable sentence, when a layman would think them a paragraph. Reading and understanding claims is important.
2. In an interview with an examiner, if you do know what you are talking about, it will be apparent to him. Do *not* "pull rank" to impress him with your long, practical experience and his "ivory tower" misunderstanding of the situation.
3. In an interview, *stick to the subject* and avoid personal angles. If you are right, *demonstrate* the device which can convince him—do not try to talk him into it.
4. If you receive a Patent Office Action with a reference to another inventor's work which appears to be a clear, "right on the button" preconception of your device, *reserve all opinions and do not accept defeat.* Study that patent. *Build his device* if you can. Take the trouble to study it thoroughly enough to detect any important differences. Remember that a *pen drawing can be deceptive.* This advice, given to the writer by Perry Pattison, years ago, turned devices which were "just like yours" into devices which had no relationship, after they were thoroughly understood. Remember: there is always time to accept defeat . . . later. As for your study: *build* the device, see it work . . . then decide.

The "Anatomy" of a Claim

As mentioned previously, the drawings and specifications are necessary to the public to understand and eventually use your invention; but the Patent Grant to you is spelled out in great detail in the *claims,* and *only* the claims tell what you have covered. The drawings may show something (to illustrate a necessary point), but unless the claims recite it, it is *not* covered.

The claims of a patent are the part in which the physical structure is described in unmistakable detail. *A claim is one sentence,* usually punctuated with innumerable commas, semicolons, colons and a single period. It must describe *YOUR* invention *ONLY* and not read on (describe) any other invention. If a single, accurate sentence cannot be written to encompass your invention, then perhaps you have no invention. The first claim which is printed in your patent, at the end of the specifications, is your generic claim—the prime, most important claim. Once that is granted, you are then allowed secondary claims for the lesser parts of the invention, or for the variations of it. For you to comprehend a claim, try to break it up into little areas so that the entire picture is built up in your mind like the parts of a jigsaw puzzle being put together. Forget that it *is* a single complex sentence—make it several simple-to-understand sentences.

When an inventor improves the device of another inventor, his claims may be allowed to read on the original invention *plus* his own contribution. This is because, alone, it would not be a functioning device; it is dependent upon the original mechanism. While nowhere in the patent is the term "dependent patent" actually used, it is most definitely a dependent patent and the inventor cannot legally make the other man's invention.

Why, then, give him a patent at all? To prevent the

original inventor from using that improvement without paying the man who improved it. A dependent patent assures the inventor of his due consideration for improving the device. No one can read a patent claim, or even a patent, and accept its statements literally. One *must* know the field to see the relationships of the various existing patents and their effects upon your own or the patent being considered. The Patent Office is full of improvement patents to help advance certain broad arts, but improving the invention of another does not give the improver the right to appropriate the original invention.

Because these situations *are* so complex, patents and legal opinions are not the province of every inventor; the attorney himself could be wrong, hence the prevalence of lawsuits with opposing sides. However, if the inventor does strive to understand the field and the procedures the learned ones follow, he will be rewarded with a better understanding of his own problems and experience fewer unnecessary troubles.

The Patent Attorneys for the Neophytes

Because of the nature of his business, the writer has had occasion to work with a score of attorneys, and has found them to be able, ethical and trustworthy. However, because of the nature of the "off the street" neophyte inventor, he often has to accept arrangements with an attorney which the professional or business inventor would never tolerate.

The inexperienced inventor wants to know, in advance, what his costs are. Since no two cases are alike; no two fields are of equal complexity and an amateur's inventions are *frequently* in need of beyond-normal assistance, a patent attorney who names a fee in advance has to be either a person who pads his bill to cover these unfair demands or a person who employs tricky shortcuts in drawing up a patent application. It is the latter practice which, unfortunately, has usually been chosen, for the neophyte seldom has enough money to pay a regular bill, much less a padded one.

Despite being a beginner, an inventor may indeed have a very important first invention, and, if he is forewarned of unethical practices, he will be better able to combat them. One such practice which carries no illegal implications, yet which is not only unethical but injurious to the inventor, is the "additional feature" which the draftsman or attorney "contributes" to the invention.

To illustrate what this is and why it is done, let us take a hypothetical case: An amateur inventor designs a toy steam shovel with a novel and ingenious mechanical action. However, he demands a low, flat price for the attorney's fee. The attorney knows that toys are a very large and heavily patented field, and that the case could demand an expensive search, a good deal of study and possibly thinking his way out of close situations. However, he suggests that the inventor could, with no additional cost, "add a novel feature" to his design . . . as the shovel picks up a load, it emits a sound like a steam engine makes, adding "realism" to the toy.

The inventor follows the suggestion. The attorney knows that the combination of the two elements will be unusual enough to make the patent application far easier; there being few if any conflicting patents may insure his getting a quick and less costly patent. However, our neophyte does not realize that any other toy maker can use his patented invention without royalties merely by leaving out the feature of the sound, which is completely unnecessary. A worthless patent!

This, and other ways of getting a quick patent (and a worthless one) are the province of the attorney catering to the one-shot, flat-price beginner. Those who advertise the bargain prices and mail-order services are considered less desirable, even by their own fellow attorneys. This does not, nor cannot mean, that *all* advertising attorneys are unethical, but it does mean that low-cost, flat-price applications may be subject to hidden dangers.

CHAPTER SIXTEEN

HOW TO PROTECT YOUR IDEA

No man can lightly toss off "wonderful ideas" in everyone else's field and have no field which is really his own. Just as a lawyer becomes a tax lawyer, or a doctor becomes a heart specialist, or an artist, a portrait painter or fashion artist, etc., just so, an idea man must operate in a restricted area where he is more knowledgeable, more creative and more ingenious than others in that area. Only then will he succeed with the marketing of his creations.

Are "Ideas" Protectable?

IF one accepts the *Random House Dictionary's* definition of "an idea" as being "a thought—a conception of the mind." Then the answer is definitely "no!" There is no government patent, copyright or other form of protection for a thought in your mind. Fortunately, ideas *can* be given certain *physical forms* which *are protectable,* and those forms are so many and so varied that a man who really is creative can usually find some form of protection for his thoughts—in their *physical* form—even when he cannot patent them.

Even commercially valuable ideas, some with a degree of inventive ingenuity, cannot be protected by patents, despite the fact that they can bring financial success to their creators. Examples: selling soft drinks in sixes instead of singly—only the carriers could qualify for patents. Selling foil, wax paper and paper towels in a single, handy rack—great idea but the *idea* cannot be covered.

Types of Protection

What is it that can be protected, and what form of patent or protection is available? There are *many* forms of protection for creative people, but each is based upon giving their creations a *physical form.* Here is a brief resumé of what our government offers to those who create.

1. For *inventions* of mechanical, chemical and electrical devices, which are deemed useful and with sufficient ingenuity . . . The United States patent.
2. For *artistic designs* of clothes, furniture, automobiles and articles of commerce and manufacture . . . the design patent.
3. For original, asexually produced plants . . . the plant patent.
4. For *artistic things,* such as songs, poems, novels, magazine or newspaper articles, cartoons, cartoon characters and countless other items . . . the copyright.
5. For *trade names* and *trademarks* in actual use . . . the long-term protection offered by trade name or trademark registry.

Despite the apparent ingenuity and importance of the "art," gag men or idea men cannot copyright a joke, gag or situation. A TV Show cannot protect the idea for a give-away program. No one can patent the *idea* of white lipstick; multi-unit sales of beer or soft drinks; false eyelashes; pureed baby foods; etc, etc. Some of those items can have their particular *form* protected, but then others may find many other ways to accomplish the same result. Moral: give your ideas physical forms which *can* be protected. Ideas, as such, are of value only to someone *in* the business.

It is hard to kill the myth of people, off the street, just walking into a big company, making a brilliant suggestion and walking out with a princely check. It just isn't being done!

Getting the Details About Copyright Protection

The writer considers it absolutely mandatory for anyone planning to write, draw, create or otherwise live by his ideas, to get the wonderful *free* booklet entitled General Information on Copyrights by sending for it to Register of Copyright, Library of Congress, Washington D.C. 20540. Unlike the mechanical patents, copyrights may be obtained without prior submission to the Copyright Office in Washington. *After* the book, or poem, or song, or novel is already printed, and *already marked "copyright,"* then a few copies of it are sent to the Copyright Office. In view of the fact that different rules apply to different artistic things, no *one* set of directions will be given here.

Copyright can cover such dissimilar artistic creations as sculpture, decorative maps, lyrics for a song, newspaper columns, even photographs. *All* copyright protection is very low in cost, and most do *not* require a patent attorney's services. Each separate category of creation is listed and the requirements for it made very simple, so protecting your work becomes far, far less costly than with mechanical patents.

When, for certain reasons, an author may desire to obtain a copyright on an *un*published work, he can do the following: Follow the directions given by the booklet General Information on Copyrights previously mentioned. On page 5 of that booklet he will find the heading, Copyright Procedure for an Unpublished Work. There he will learn of what is known as a "statutory copyright," offering just that type of protection. The cost of such registry is only $6.

How You Can Publish Your Own Book . . . Cheaply

When the Copyright Office specifies "published work," they do *not* detail how large a printing would be considered "published." Therefore, with very inexpensive offset printing, a dozen or so can be printed and handbound with any one of half a dozen ready-to-use bindings, or by arranging the pages in numerical order and gluing the back edges together to make a book or booklet. The writer has at times, stacked up a dozen or so sets of pages, all properly arranged, and with a latex-based glue like Sobo or Elmer's Glue, covered the back edges, let them dry and then slit them apart at the right places with a large knife. It is best to place a substantial weight on the stack before gluing.

For smaller, limited runs, Xerox has a process which duplicates a book or booklet; records it on film in sequence and, from that point on, makes duplicates from the film. They give you the film for future use. Having several copies makes submissions to agents and publishers far more convenient, and still allows you to keep a few.

CHAPTER SEVENTEEN

HOW TO GET YOURSELF A GOOD CONTRACT TO PROTECT YOUR IDEA

If the purpose of a written contract were condensed into a single sentence, it might read: "An honorable statement of intentions, by two or more parties, to prevent misunderstanding and insure performance."

THUS, the understanding of the day is placed in written form so that tomorrow, next month or ten years hence, that understanding remains unchanged. The passing of time causes even the most well-intentioned people to think they said things which they did not. Contracts are instruments of good intent—beware the person who avoids putting a verbal agreement in writing. In good times, you may get what was promised; in bad times, he is free to give you nothing . . . and you cannot hold him.

Why Verbal Agreements Are Unbusinesslike

Gone are the days when "gentlemen's agreements" were all that were required—a good handshake and it was a deal. That was fiction even then, for men still dropped dead of heart attacks, were killed in wars and accidents and left situations which their survivors found had no legal basis for existence. The movie magnate, Sam Goldwyn, put his comic touch to a most clearsighted bit of business philosophy when he said, "Verbal agreements aren't worth the paper they're written on."

A great danger in our age exists in the growing tendency of publicly held companies to merge for survival. When a corporation is bought, all friendly considerations and verbal understandings are considered nonexistent. They carry no obligation for the new owners. With the vast number of mergers now taking place, *only* your written contracts and provisions will be binding upon the newcomers.

People who were shortsighted enough to make unethical arrangements to avoid the paying of taxes, find themselves in a position where the new management not only doesn't have to honor arrangement, but the parties to the agreement can't resort to the law because they were doing something illegal and dishonest. A good, clean-cut contract without guile or subterfuge is, indeed, the best for all.

The Essential Elements of a Royalty Agreement

The writer has assembled, in this chapter, all of the essentials which a good Royalty Agreement should have. If other factors must be added for specific cases, add them, but do not do so by eliminating those proposed herein. It took the lifetime of many able people to bring these elements into so simple and efficient a contract; don't leave out any element because "that's obvious" or "naturally, we all assume that!" Here is an example to remember.

A newly married couple go to their honeymoon motel and the groom registers. He is flustered and inexperienced. The clerk asks him "Who is that girl with you?" He replies, "My wife, of course!" "Well," says the clerk, "you forgot to mention her on the registry card. You just signed in for yourself." Perhaps, like the groom, you too overlooked a point as obvious as the wife who stood beside you. It is just those obvious, but important points which we must check on, through the use of the following list.

Essential Elements of a Royalty Agreement

1. The title of the document: *Agreement* or *Contract.*
2. The date the Contract was signed; it can be before or after the terms, but must be definitely established.
3. The *full* name of each of the contracting parties.
4. A shortened designation of each full name, so the long version does not have to be repeated each time (see sample Agreement which follows).
5. The correct address of each party.
6. A statement that the parties are individuals, companies or corporations.
7. A brief statement establishing the type of business each party conducts.
8. A statement of the general intent of the parties toward each other, and how they shall work together under the following numbered paragraphs.
9. A statement of the payment, remuneration or obligation which the licensee will provide, this being the reason for the entire contract.
10. Acceptance of the financial consideration by the licensor (inventor) with a statement by him of what he will grant and do for that money, in broad, general terms.
11. Several separate paragraphs stating in detail what each party will do, dividing them so they can be numbered . . . thus making them easier to refer to.
12. A clear, simple specifying of the percentage of royalty to be paid, and whether that percentage applies to the manufacturer's price or the retail price. A statement of what may be deducted, such as trade discounts, freight, etc.

13. A guaranteed, minimum royalty, with a choice of methods of payment: straight royalty earned or license cancelled; privilege of licensee to make up in money for deficit in royalties earned.
14. Specific times when royalty payments are due, stating both the check to be delivered and the accounting for the sales and percentages, discounts and deductions. Some Agreements state that the licensor retains the right to examine the books, if necessary. This is *not* a must—but optional.
15. The terms under which either party may terminate the Agreement; when notice must be given; how delivered; the time allowed between notice and termination.
16. The disposition of the component parts of the Agreement following termination: what to do about patents, licenses, sublicensees (if any); customers already served; prospects already solicited but materializing after the termination date; stocks of goods made but not yet sold; equipment now held unusable and, finally, royalties to be paid on goods sold and shipped after termination.
17. Summary of Agreement, stating that the number of paragraphs which comprise the Agreement are the entire understanding between the parties, and that no other provisions, verbal or implied, shall be considered pertinent. State that all changes which become necessary must be placed in writing and signed by both parties, to become part of this Agreement.
18. Provide places for both company names, and space beneath each for an officer of the company or individual to sign, adding his title, if any, and the date signed.

Because Agreements and Contracts can have a formidable appearance and be "heavy reading" for all but the attorneys, the writer has adopted the style shown on the succeeding pages, which breaks the sentences up into easy-to-read, related thoughts, thus making for legal documents which even the average layman can understand clearly.

The following Agreement is a sample of a hypothetical Royalty Contract between an inventor of shipping boxes and a box manufacturer, which illustrates how the simplified style is to be used. Note the simple, logical passage from introduction of parties, through statement of intent, into specific terms and obligations, into acceptance by signatures.

AGREEMENT

AGREEMENT

made this ________ day of ________, 1973, between Consolidated Corrugated Case Corporation, a Delaware Corporation with headquarters at 25-32 44th Street, Long Island City, New York, hereinafter referred to as "Consolidated."

and

Arthur K. Valentine, an independent inventor residing at 86-40 101st Street, Kew Gardens, L.I. New York, hereinafter referred to as "Valentine."

WHEREAS

Valentine is an inventor of corrugated shipping boxes and paperboard constructions,

and

WHEREAS

Valentine has been granted United States Patent No. 3, 195, 734 covering a partitioned shipping box for glass bottles,

and

WHEREAS

Consolidated is engaged in the selling and manufacturing of corrugated shipping boxes and paperboard constructions,

NOW, THEREFORE,

Valentine and Consolidated have agreed to work together, for their mutual good, to promote the sale and use of Valentine's partitioned boxes, under the following terms:

1. Consolidated agrees to pay Valentine, a royalty of 5% on the gross amount of all sales made by Consolidated, of partitioned shipping boxes, covered by the claims of his United States Patent No. 3, 195, 734.
2. Consolidated further agrees to pay an advance against Valentine's royalties, the sum of $2,500, said sum to be a minimum guaranteed royalty, and, in the event of default by Consolidated, not refundable by Valentine.

3. For the considerations described in paragraphs 1 and 2, herein, Valentine hereby agrees to grant to Consolidated, an exclusive license to make, use and sell boxes of said patented design, within a territory of a 250-mile radius of New York City.
4. Valentine further agrees to design and build boxes of his partitioned construction, for whatever specific containers Consolidated will bring to him. For this service, Valentine will receive a further nonrefundable advance against royalties, of $100 per project.
5. The royalties which Consolidated will pay, will be subject to the standard trade discount of 1% for payment by customers within 10 days, and such advances as customers may demand for freight charges, prepaid trucking, etc. No royalties will be paid on returns or damaged goods.
6. When the royalties accruing to Valentine during any single calendar year exceed the sum advanced to him by $300, then all further excess will be paid to him, on or about the 10th of every month, for goods shipped the previous month. At the end of each calendar year, the withheld amount of $300 for that year will be paid on or about January 10th of the following year.
7. If either party to this Agreement desires to terminate it, then written notice to that effect, must be delivered by registered mail, to the other party, at least 90 days before the end of that year; termination to take place at the year's end.
8. In the event of termination, Consolidated will retain the right to finish all work started before the year's end, and fill reorders from customers whose accounts were obtained during the Agreement, said reorders to be so respected for a period of two years past termination.
9. The royalty payments of 5% will continue throughout all post-termination business, but no advances will be paid by Consolidated.
10. Valentine agrees that any new licensee must respect Consolidated's rights on post-termination business, and that said licensee will not bid on nor accept Consolidated's customers' business until the full two-year period is past.
11. If the volume of business sold by Consolidated equals or exceeds the sum of $200,000 annually, then Valen-

tine may not exercise his right to terminate during that year.

12. In the event of disagreement not capable of settlement by simple discussion, both parties hereby agree to submit their differences to the impartial judgment of the American Arbitration Association of New York City, and will abide by their decision.

The 12 paragraphs of the Agreement comprise the entire understanding between the parties, and no other terms, verbal or implied, will be considered pertinent. To be valid, changes must be placed in writing and signed by both parties.

Accepted for "Consolidated" Consolidated Corrugated Case Corporation	Accepted for "Valentine" Arthur K. Valentine
______________________________ Title: Date:	____________________ Date:

CHAPTER EIGHTEEN

IDEAS FOR BUSINESS/PROFITS WITHOUT PROTECTION

Despite the lack of patent protection and the obvious dangers of having one's ideas legally appropriated by competitors, imaginative businessmen have created long-lived, highly profitable, admirably ingenious business ideas, which seldom are equalled financially by inventors.

A specific field is rarely open to outside idea men. In the writer's experience, the good ideas are created either by the heads of the business or the top executives employed for their creative or promotional talents.

Analyze the following "success stories," and note *who* "put them over," *who* was in a position to benefit from the ideas. In every case, the profits went to those who transformed the ideas into actions . . . *selling* the merchandise or services themselves, *not* trying to sell the ideas for cash or royalties, leaving the promotion to others. Business ideas are *not* creating new mechanics nor new merchandise, but, rather, *filling an established need* with a new approach, a new "twist" or with commercial ingenuity applied to sales problems.

New Keys to Business–New Business in New Keys

When the Cole National Corporation opened its chain of key-making departments in the Sears Stores, and shortly thereafter in other chain stores, it had made available to the buying public a new and convenient place in which people could buy duplicate keys. There was nothing new in the simple, machine key duplicator; many hardware stores had them, but women (and men, too) seldom visited hardware stores; they did visit the chain stores often. Availability is a most powerful factor in selling, and that worked in their favor.

While not rating the appellative "invention," Cole National introduced keys of tempered aluminum instead of steel, and the keys were lighter in the pocket. Then, with aluminum in place of steel, they could color them with anodizing so that keys could be "color coded" . . . a red key may be your front door key; a blue one the key to your office; a yellow one a key to your locker. The simple convenience of recognizing color instead of puzzling out shape, was appealing to the public, and they bought it. Hundreds of these key concessions in the chain stores prospered because of this new and enterprising way of handling an old, stuck-in-a-corner business.

Next, Cole National took the eyeglass business, and put it on a simpler, more businesslike basis than the typical optician's "expensive looking" store (an attraction to the many people who had limited means). They figured out that each eyeglass counter in the chainstores need only have a space where the attendant could examine the person's eyes, using a machine which did not require an oculist to operate. The people were well trained in their jobs but did not treat eye diseases; they merely fitted customers to well-suited lenses and comfortable frames.

Then, in a central location, a good-size lens grinding facility and eyeglass factory received the accurate prescriptions from the hundreds of chain stores, and filled them efficiently with top-quality help, each one being kept far more productive than any single shop could ever keep them. Thus with examining facilities only in the expensive chain locations, and concentrated lens grinding in the less-expensive factory location, a prosperous business was founded. Besides the concept, it took much work, much thinking and the ability to organize people. It also took the know-how of good financing to swing it. Without the concept—the *idea*— it is doubtful that such an enterprise could be created; with it, all the organized, able,ready-to-work staff knew where to go and what to do.

Profits in Pleasures

In a suburban section of the north shore of Long Island, a group of apartment houses were built, to house typical, family-type American people with their children. The houses had one special feature, however, which surely would be an attraction to almost any one of the hundreds of such housing developments throughout the nation, but sadly, few have it. It is a handsome swimming pool. This handsome pool, with a broad terrace all around it, is a concession, run by a most knowledgeable, psychologically astute, businessman.

The pool has been turned into a most enjoyable tenants' country club, and it has solved the problem of the typical city dweller . . . What to do with the hot summers, even if he does have his two- or three-week vacation? What to do with restless kids, a hot and uncomfortable (or even lonely) wife? With the pool, instead of just leaving things to chance, this smart concessionaire organized it so that it serves *all* the tenants' needs, each at different times.

A Smarter Way to Run Things

Colorful tables with umbrellas and comfortable chairs surround the pool on the terrace. In the daytime, sandwiches are served with soft drinks there; in the evening, dinners are served. The pleasure of eating at this "country club" pool is a high spot in the tenants' day. Later, dancing is allowed, and with recorded music, they have such an innovation as "Teen-Agers Night," then on some other night, "Pre-Teen Night." Saturday nights are always reserved for the adults, and they dress accordingly.

Besides the pleasure of the food and pool, there is the fun of mixing with neighbors, all about the same age and with kindred interests. For these pleasures, the concessionaire charges a few hundred dollars a year as a membership fee, then sells the drinks, food, candy, cigarettes and other items. It is indeed a "happy business." The pool runs through June, July, August and September, and during that time requires a small crew of helpers. During the other eight months, the concessionaire runs things as the occasion requires.

Such a project as this can be added to apartment house projects later, when space permits, but, obviously, getting in on the building project in the planning stage allows for better facilities.

A Jewel of an Idea

One of America's most enterprising merchandisers is the Jewel Company, formerly known for generations as the Jewel Tea Co. In one division of this company, they have a huge chain of supermarkets, mostly in the midwest. In another division—the one which now concerns us—they sell directly to the consumer, going right to her house to do so.

This is *not* to be confused with the so-called door-to-door canvassers, for the *idea* which the heads of the Jewel Company conceived, lifted their men out of the class of unwelcome peddlers and earned them a perpetual, periodic welcome from their customers. It was most cleverly figured out to *serve the customer well . . . not* just to make a profit for the company. They were wise enough to know that only by serving the customer with what *she* wanted, could they continue to get what *they* wanted . . . steady business.

Every two weeks, the Jewel man drives his delivery truck to *his* neighborhood, where he has earned his place as a faithful, dependable tradesman. He has a *new* attractive folder each time, showing the items for sale . . . *staples only*—no perishables! His prices are usually slightly lower than the supermarket's, and the customer does not have to lug home heavy boxes of cleanser, soap powder, toilet tissue, paper towels, furniture waxes, laundry bleaches and all the staple household "musts." He always has some special item at a very low price. The customer, in the comfort of her home, or on her porch or lawn, sits down and studies the attractive "menu" or list. Colorful pictures and big text lighten the reading. She doesn't need glasses.

The Jewel man takes her order for delivery *two weeks hence. If* some special item is needed in a hurry, he may drop it off for her sooner. At this point, he collects for the last order as he delivers it, and tells her about the *big special*—each week it's different . . . sometimes an electric fan (in summer of course) or a new light lawn chair . . . an Easter special of a cute suit for a little boy and a dress for a little girl (inexpensive and beautiful).

Because of their enormous buying power, they get up a special for each two-week visit: anything from plastic mixing bowls to an electric mixer, from a smart housedress to an electric fry pan. The Jewel salesmen turn in a remarkable sales record, and the smart merchandisers who worked out this "welcome re-entry plan" perfected it over the years. To keep it alive, they have assembled a team of imaginative men to prepare new items–each appropriate to a time or season–and present them in the most attractive and quick-to-display manner. This *is* a business idea worthy of study.

Thousands vs. Dozens

When you hear the story of certain successes, you cannot help but say "That man deserved his success!" The story of one Jerry Lewis (no! *not* the comedian) and his brother Bob is just such a story. Jerry had a very clever idea for an insulated cup which kept coffee hot a long time. In the form of a soup bowl, it served the same purpose. It was made up of two pieces: an inner shell of opaque heat-resistant plastic and an outer shell of clear heat-resistant plastic. The air space between was the insulation.

Into the hollow insulating space, Jerry Lewis inserted colorful grass cloth or monk's cloth fillers, which, seen through the transparent outer shell, gave great beauty to the coffee cups and bowls. When the coffee cups (or more accurately, mugs) were laid to rest with the hot liquid in them, on polished furniture, they left no disfiguring ring, as an ordinary china cup would. When this was established, he made cold-drink glasses, too, and these left no telltale rings. Housewives loved them for their beauty as well as their ability to avoid leaving wet rings on furniture.

Had Jerry Lewis been a typical businessman, he would have sold his cold-drink "glasses" and his hot-drink mugs to the gift shops, department stores and housewares shops, where they would have ordered a few dozen at a time and the business would have been of modest size. However, he was not a typical businessman; he was a most *imaginative* businessman. He found a way to sell his insulated mugs, bowls and pitchers, in huge quantities, by one really clever concept.

Mr. Lewis had learned that the giant dairies in the suburban areas, delivered milk daily, and the same milkman sold such other dairy items as cottage cheese, whipping cream, cream cheese, coffee cream and many other items. Because milk itself is so necessary to the nations's young, the government controls its price, and milk, therefore, is a very low-profit product. The government, however, does not regulate the prices of less vital products, leaving them to the competitive situation to hold a reasonable price level.

Seeing this situation as an opportunity for his wares, Jerry Lewis sold the big dairies on the idea of filling his coffee mugs and soup bowls with their cottage cheese, and with a smart-looking plastic lid on them, selling the housewife the beautiful insulated mugs and bowls at a low cost. He had achieved distribution of his wares in the better stores, so the homemakers got to know the real value of the insulated wares. With the huge orders the large dairies placed, the price came down very low; also, the dairies were trying to get their customers into the cottage cheese habit, not just make a few cents on the mugs and bowls.

After a woman realized how nice it would be to have a whole set of the mugs or bowls, she bought cottage cheese regularly and soon had a set. Then she would send requests to Mr. Lewis, direct, to buy such items as insulated trays and pitchers, in matching colors, to complete her set. Thus he turned his product into a large-volume product, and soon benefited by the increased business. He opened up a brand-new factory in Brooklyn and ran three shifts a day to keep up with his orders. Here, indeed, is proof that business idea men are truly creative! He is not to be judged by how many ideas he had, but the worth of each, and this was an inspiration to other businessmen to see opportunities in their own backyards.

Selling Imagination Instead of Insurance

When Equitable Life Assurance Society trains their new salesmen to sell insurance, they do not let them blunder about cold, nor do they let them go out and think up their own ways of selling. Insurance is a difficult thing to sell; it would seem that everyone has an uncle, or cousin, or brother to whom they feel they "owe" their insurance business. Selling without an idea is indeed difficult.

One plan that Equitable has put into operation these past few years is scientifically figured out to benefit everyone and hurt no one, but it must be thought out in such a way that only a precise, planned campaign could set the stage for the selling. With so many middle-aged people wondering how they can possibly afford to send their children to college at today's level of tuition cost and living expenses, Equitable sales managers thought that these problems were actually golden opportunities for their men.

Suburban neighborhoods were chosen by the criterion of when they had been built, with the neighborhoods between 16 and 20 years old considered ideal for the men's efforts. These were the people who had children preparing for college. These were the people who saved and scrimped to make higher education possible for their offspring. The Equitable men were trained to send out letters to all the people in their chosen neighborhoods, saying that Equitable had a plan to furnish all the funds necessary to send their children to college, and it would not cost them anything; they merely had to inquire by letter or phone.

The inquiries from worried parents were indeed plentiful. The Equitable man would visit the prospect in the evening when the man and wife could be present. He told them a story approximating this: When you bought your house 15 or 20 years ago, you paid about $12,000 to $15,000 for it—today, it is worth well over double that amount. In those 15 or 20 years, you have either paid off your mortgage completely or brought it to a very low figure.

Now, because of increased real estate values, Equitable will give you a mortgage from $5,000 to $10,000 above what you now have. They will sell you a policy with it so that if you die before the mortgage is paid up, the policy will complete the payments automatically and the house will once more be clear. The money will provide years of education for your children. A further low-cost policy will insure your home against fire, or any natural destructive cause—from cyclone to flood.

Thus, with no selling of insurance, as such, Equitable got a mortgage investment, a life policy and a fire and flood policy. Also, the children received their college education so that they could earn more throughout their lives. That *is* good thinking, and another example of how ideas can improve any business . . . from insurance to fried chicken, from coffee mugs to swimming pools.

Business is filled to the brim with opportunities for those who have trained themselves to look for them. Just being aware of the situations about you alerts you to the good things, which, had you not been so alerted, you would have missed completely.

One caution, however: let your experience in your own business guide you. Ideas for the other fellow may look good because you are not close enough to the field to see the pitfalls. This is a world for courageous people, but courage to act without judgment to guide is a situation fraught with danger.

It takes *two* forms of ability to succeed with any business idea: first, the ability to create based upon judgment and experience; second, the ability to act upon the idea and exploit it commercially. Business *ideas* are not marketable to others, just as unpatented inventions would not be, but for those who sell merchandise and services, government protection is unnecessary.